Geology of the Invermoriston district

The Invermoriston district lies at the heart of the Caledonian orogenic belt in Scotland and astride the Great Glen Fault. To the north-west of the fault are the Neoproterozoic metasedimentary rocks of the Moine Supergroup, originally sandstones (now psammites), with some siltstones and mudstones (now semipelites). These rocks, belonging to the Glenfinnan and Loch Eil groups, were deformed and metamorphosed, initially during late Precambrian nappe-forming events, and later in middle Palaeozoic times. Early in their tectonothermal history the Moine rocks were intruded by granitic sheets and sills of basic rocks. During the early metamorphism, the former were transformed into a granitic orthogneiss while the metasedimentary rocks were variably migmatised. Alkaline igneous activity marks the onset of the Palaeozoic tectonothermal event, with the intrusion of an unusual suite of metabasic rocks. Evidence of tectonic reworking within the district during this event is minimal. Following the metamorphic peak, the district was extensively veined by granite and pegmatite, which were subsequently deformed.

South-east of the Great Glen Fault, the Grampian Group of the Central Highlands consists mostly of a thick succession of predominantly psammitic rocks, with some semipelites. A zone of ductile shearing separates the Grampian Group from an underlying coarse clastic psammite sequence. The latter is of uncertain stratigraphical affinity, but has some similarity to parts of the Moine succession of the Northern Highlands. Apparent disparities in metamorphic grade across the shear zone suggest substantial crustal transport. The tectonothermal history of the Central Highland terrain is similar to that of the Northern Highlands, but of uncertain age.

The Caledonian orogenic cycle concluded with extensive calc-alkaline plutonic magmatism, during the late Silurian to early Devonian. Although the intrusive histories on both sides of the Great Glen Fault are broadly similar, direct correlation is not possible due to lateral sinistral movement on the fault. There followed a period of rapid crustal uplift and erosion in Devonian times, with the deposition of the Old Red Sandstone red-bed sequences in fault-controlled intermontane basins. The sedimentary rocks were involved in tectonic events related to continuing movement in the Great Glen Fault system. A hiatus in tectonic activity ensued, until Permo-Carboniferous times, with the instigation of new fault systems and reactivation of the Great Glen Fault. These were accompanied by minor igneous activity, and the emplacement of alkali lamprophyre dykes.

The pre-Quaternary landscape was modified by glaciations throughout the Pleistocene. The last full glaciation, the Main Late Devensian, removed most products of previous glaciations and weathered bedrock. The subsequent retreat of the ice was reversed between about 11 000 and 10 000 years ago, during the Loch Lomond Stadial, when glaciers once more occupied parts of the main valleys.

Following the final retreat of the ice, the valley floors remained infilled with glacial deposits, morainic debris and glaciofluvial sands and gravels, which were subsequently reworked by the rivers. Peat has accumulated extensively since the final glaciation, particularly on the higher ground.

Cover photograph
Loch Tarff, view south-westwards across hummocky glacial deposits to Borlum Hill and the Great Glen.
(Photographer: T S Bain)

Flat-lying beds on the long limb of asymetrical D$_2$ folds deforming interbedded psammite and semipelite, Achnaconeran Striped Formation. River Moriston.

BRITISH GEOLOGICAL SURVEY

F MAY and
A J HIGHTON

Geology of the Invermoriston district

Memoir for 1:50 000 Geological Sheet 73W
(Scotland)

CONTRIBUTORS

Geology
G C Clark

Geophysics
B C Chacksfield

London: The Stationery Office   1997

First published 1997

ISBN 0 11 884532 2

Bibliographical reference

MAY, F, and HIGHTON, A J. 1997.   Geology of the Invermoriston district.   *Memoir of the British Geological Survey*, Sheet 73W (Scotland).

Authors

F May, BSc, PhD, DIC, CGeol
formerly British Geological Survey, Edinburgh

A J Highton, BSc, PhD, CGeol
British Geological Survey, Edinburgh

Contributors

G C Clark, BSc, PhD
formerly British Geological Survey, Edinburgh

B C Chacksfield, BSc
British Geological Survey, Keyworth

Other publications of the Survey dealing with this district and adjoining areas

BOOKS

British Regional Geology
The Northern Highlands, 4th edition, 1989
The Grampian Highlands, 4th edition, 1995.

Memoirs
The geology of the country round Beauly and Inverness (Sheet 83), 1914
Geology of the Glen Affric district (Sheet 72E), 1992
Geology of the Glen Roy district (Sheet 63W), 1997

Technical Reports
The Old Red Sandstone of the Mealfuarvonie Outlier, west of Loch Ness, Inverness-shire, No. 83/7, 1983

MAPS and ATLAS

1:625 000
United Kingdom (North Sheet)
Solid geology, 1979
Quaternary geology, 1977
Bouguer anomaly, 1981
Aeromagnetic anomaly, 1972

1:250 000
Great Glen Sheet(57N 06W) Solid geology, 1989
Geochemical Atlas Great Glen, 1987

1:63 360
Sheet 83 (Inverness) Solid and Drift, 1914

1:50 000
Sheet 63W (Glen Roy) Solid and Drift, 1995
Sheet 72E (Glen Affric) Solid, 1986
Sheet 73W (Invermoriston) Solid, 1993
Sheet 73E (Foyers) Solid, 1996

Printed in the UK for The Stationery Office
J27602   C6   11/97

CONTENTS

FIGURES

PLATES

TABLES

PREFACE

The Invermoriston district lies at the heart of the Scottish Highlands. The district consists of rugged mountains and peat-covered moorlands, cut by the flat-bottomed valleys of the Great Glen, Glen Moriston and Strathglass. Much of the area is sparsely populated, with habitation largely concentrated within the valleys. Fort Augustus, with its famous flight of locks on the Caledonian Canal, is the largest centre of population. Lying astride the Great Glen, the district is one of the most popular areas for tourists in the British Isles. The diverse range of visitors attracted to the district attests to the particular natural beauty and vistas of this part of the Scottish Highlands, an attraction enhanced by the 'enigma' of Loch Ness.

Understanding of the geology of the UK is essential both to proper conservation and development. It is particularly relevant to exploration for resources, the avoidance of hazards, and sensible land-use planning. In recognition of this, the British Geological Survey is funded by central Government to improve our understanding of the three-dimensional geology of the UK national domain through a programme of data collection, interpretation, publication and archiving. One aim of this programme is to ensure coverage of the UK land area by modern 1:50 000 geological maps, mostly with explanatory memoirs, by the year 2005. This memoir on the Invermoriston district of Scotland is part of the output from that programme. It is based on the primary geological survey undertaken between 1975 and 1985.

The metamorphic rocks of the district form part of the Caledonian Mountain belt, although the landforms we see today are the result of a chain of events spanning over one thousand million years. The period of mountain building resulted from collision between the Laurentian and Gondwanaland continental masses. This commenced during the late Precambrian (about 900 million years ago) and continued to middle Palaeozoic times (about 430 million years ago). During this period, the sedimentary rocks of the Precambrian sequence were involved in complex folding and metamorphic processes, and intrusion by a wide variety of igneous rocks. There followed a period of uplift and erosion, mainly during Devonian times (about 410–360 million years ago), exposing rocks which had been buried at depths up to 20 km below the mountain tops; there was also deposition of Old Red Sandstone sediments in intermontane basins. However, the diverse landscape as we know it today owes much to the modification of the pre-existing mountainous topography by Quaternary glaciation. This resulted in erosion of the mountains, smoothing them off, and widening of the valleys. The final retreat of the glaciers left a thin, but significant, cover of glacial and periglacial deposits, particularly on the lower slopes and in the valleys.

No significant amounts of metalliferous minerals are known, but there are some substantial deposits of sand and gravel and a large potential resource of hard-rock aggregate. However, the main resource of the area is its landscape, which will continue to attract many visitors as part of our natural heritage. In 1958 I was fortunate to be involved in a student expedition to Glen Affric in the western part of the district. I have never forgotten that experience, or the beauty of the area. At that time there was no map or memoir available, and we were much on our own in attempting (not entirely successfully) to sort out some very complex geology. This memoir obviously provides a new understanding of the geology of this part of the Scottish Highlands. I hope it will also provide a fuller appreciation of the beauty of the district and help to ensure its careful conservation for future generations.

Peter J Cook, DSc, CGeol, FGS
Director

British Geological Survey
Kingsley Durham Centre
Keyworth
Nottingham
NG12 5GG

NOTES

Throughout this memoir the word 'district' means the area encompassed by the 1:50 000 Geological Sheet 73W (Invermoriston).

Figures in square brackets are National Grid references; all lie within 100 km square NH.

Five-figure numbers preceded by S refer to thin sections held by the BGS in the Scottish Collection at Murchison House, Edinburgh.

ACKNOWLEDGEMENTS

In this memoir, Chapters 2 and 8 were compiled and written by Dr May and include an account of the metamorphism of the Moine Supergroup by Dr Highton. Dr G C Clark presents a compilation of the geology of the Grampian Group in Chapter 3 derived mainly from published sources. Dr Clark also contributed to Chapter 4, which examines rocks of uncertain stratigraphical position. This chapter was compiled and written by Dr Highton, and based partly on the thesis work of L M Parson (formely of the University of Liverpool) and the work of Dr D I Smith. Chapter 5 describing the pre-Silurian igneous history of the district was written by Drs Highton and May; while in Chapter 7, Dr Highton examines the Siluro-Devonian and Permo-Carboniferous magmatism in the district. Aspects of Old Red Sandstone geology, compiled by Dr May, derive from the mapping of Dr W Mykura and a detailed description and interpretation of the conditions of deposition and the palynology have been provided in Mykura and Owens (1983). Chapter 9, Drs Highton and May describe the Quaternary geology of the district; this is based on mapping of the landforms and deposits during the primary survey, but encompasses the work of Sissons (1977) and Firth (1984). In Chapter 10, Mr B C Chacksfield provides a description of the geophysical characteristics of the district. Dr Highton, with a contribution by Dr May, examines the economic resources of the Invermoriston district in Chapter 11.

HISTORY OF SURVEY OF THE INVERMORISTON SHEET

The district lies partly within the region north of the Great Glen in which no detailed geological mapping had taken place prior to 1960. A systematic 1:10 560 and 1:10 000 survey was undertaken by the Geological Survey in the period 1977 to 1985. The survey of the Northern Highland area was conducted by Drs D J Fettes, A J Highton, F May, J R Mendum, W Mykura, J D Peacock, N M S Rock and C G Smith, while Dr D I Smith was concerned with the area to the south-east of the Great Glen (Appendix 1). Mr G S Johnstone and Dr W Mykura were the Regional Geologists during this period. No Drift edition of 1:50 000 Sheet 73W is available due to the absence of data on the Quaternary to the south of the Great Glen, which remains to be surveyed.

Figure 1 Physiographical and locality map of the Invermoriston district.

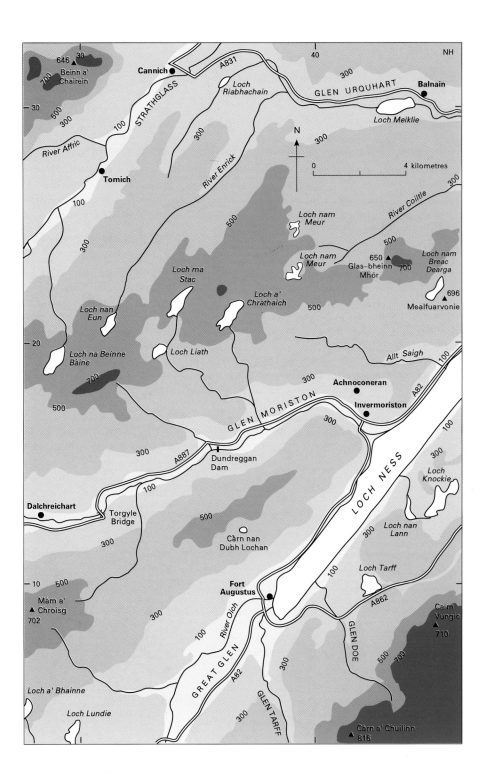

ONE

Introduction

The area covered by Sheet 73W, Invermoriston, is one of varied relief which includes parts of the Great Glen and Loch Ness (Figure 1). Glacial erosion has considerably modified an earlier landscape and the principal valleys, notably Strathglass, Glen Urquhart and Glen Moriston, have been enlarged and are now wide and flat bottomed. Loch Ness has been greatly deepened. Several hills to the north-west of the Great Glen rise to 600 m above OD and a large area between Glen Urquhart and Glen Moriston has a topography characterised by rocky hillocks and numerous lochans produced by the flow of an ice-sheet. Peaks rising to 800 m above OD lie to the south-east of the Great Glen. Solid rock is generally well exposed, although blanket peat and other Drift deposits cover much of the lower parts of the District.

Most of the district is sparsely populated, and the inhabitants are concentrated near Fort Augustus and Invermoriston. The main occupational activities are connected to tourism or employment on the large agricultural and sporting estates of the area. Forestry is also important and hydro-electric power is generated at Dundreggan in Glen Moriston and at Fasnakyle in Strathglass.

REGIONAL SETTING AND SUMMARY OF THE GEOLOGY

The Invermoriston district is divided into two parts by the Great Glen Fault Zone (Figure 2). The area to the north-west of the fault is mainly underlain by metasedimentary rocks of the Moine Supergroup, a late Proterozoic succession that was deformed and metamorphosed during the Precambrian and Ordovician events of the Caledonian Orogeny. The margin of the Caledonian orogenic belt lies about 50 km to the west. There, it is marked by the Moine Thrust, a zone of WNW-directed thrusts along which the rocks of the orogen were transported on to the foreland Lewisian gneisses and Torridonian sandstones. Lewisian rocks formed the basement upon which the Moine succession was deposited. The occurrence of Lewisian rocks within the Moine outcrop is restricted to either the cores of major regional-scale folds or tectonic slices incorporated by ductile thrusting. Lewisian rocks do not outcrop within the Invermoriston district, although they occur in the adjacent districts of Glen Affric (Sheet 72E) to the west and Scardroy (Sheet 82E) to the north-west.

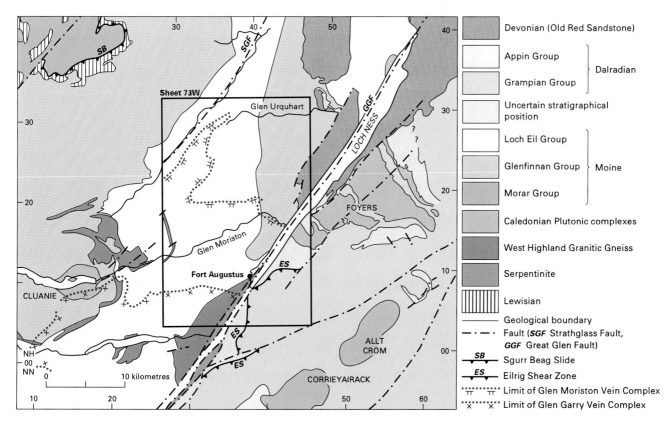

Figure 2 Regional geological setting of the Invermoriston district.

As a consequence of the general lack of good marker beds in the Moine, and the complex tectonothermal history that the rocks have suffered, neither the stratigraphy nor the structure is yet fully understood on a regional scale. It has been possible, however, to extend the stratigraphical–structural subdivisions recognised farther south (Strachan et al., 1988) into Sheet 73W. Sedimentary structures are locally preserved in the psammites allowing the reconstruction of a tentative stratigraphy for much of the Moine Supergroup within the Invermoriston district. The Moine rocks were originally deposited as clastic sediments but subsequent amphibolite-facies metamorphism has converted the sandstones into granular psammites, the shales into schistose to gneissose pelites and the sandy shales or siltstones into schistose to gneissose semipelites. Diagenetic concretions are now represented by pods of calc-silicate rock.

The oldest rocks exposed, the Glenfinnan Group, consists of mappable units of pelite, psammite, quartzite and rhythmically interbedded pelite, semipelite and psammite (striped schists). The broad characteristics of the different lithological units seems most consistent with deposition in a deep-water marine environment. The striped schists may be distal turbidites. The dominantly psammitic Upper Garry Psammite (Loch Eil Group) is probably a shallow-marine shelf deposit and the existence of herring-bone bedding indicates that parts of the succession were deposited in a regime subject to tidal activity (Strachan et al., 1988). The Glenfinnan and Loch Eil groups are linked by a sedimentary transition zone and there is evidence that the Loch Eil Group passes laterally, in a west to east direction, into rocks (Achnaconeran Formation) which are correlated with the Glenfinnan Group.

Three major tectonic events (D_1, D_2 and D_3) have affected the Moine. The disposition of the outcrops over a large part of the district is controlled by a major NNE-trending D_2 synform. The structure of the Glenfinnan Group in the north-west of the district is more complex, with interference fold patterns and appreciable slip along some of the lithological boundaries. These complexities have been interpreted as the result of late Caledonian (D_3) folds overprinting earlier, probably Precambrian, structures (Roberts and Harris, 1983). The central part of the district appears to have largely escaped the late Caledonian events, the major synform being the result of D_2 deformation which may be pre-Caledonian.

The area to the south-east of the Great Glen Fault Zone is also mainly underlain by late Proterozoic metasedimentary rocks (Figure 2) but correlation across the fault is uncertain. A narrow fault-bounded strip of impure quartzite and psammite extends from Fort Augustus to the south edge of the district. The fabric of these rocks is disturbed and, in places, destroyed by cataclasis associated with movement on the Great Glen Fault. They are tentatively correlated with the Moine Supergroup.

Most of the remaining metasedimentary rocks to the south-east of the fault are assigned to the Dalradian Supergroup. Grampian Group sediments infilled an early basin prior to regional subsidence which heralded deposition of the main Dalradian sequence. The Grampian Group comprises a thick succession of psammites, originally deposited as turbidites, overlain by shallow-water deposits, now represented by psammites and semipelites. The basal part of the overlying Appin Group, which outcrops in the Glen Roy district (Sheet 63W) to the south, consists of shallow marine deposits of quartzite and pelite laid down during a transgression caused by regional subsidence.

The original basement of the Dalradian Supergroup is not seen. A zone of ductile thrusting, the Eilrig Shear Zone, separates the Grampian Group succession from an underlying sequence of rocks of uncertain stratigraphical affinity, the Glen Buck Pebbly Psammite Formation.

Early recumbent folds (nappes), overturned towards the north-west, affect the Dalradian rocks to the south, in the Glen Roy district (Sheet 63W). These structures are overprinted by upright NE-trending folds which are more widely developed. Ductile shearing accompanied the folding. Thrusting towards the north-west along the Eilrig Shear Zone coincides with the peak of regional metamorphism. This reached middle amphibolite facies in the Grampian Group, but only greenschist or lower amphibolite facies in the underlying Glen Buck Pebbly Psammite Formation.

The igneous intrusions in the district have a wide range of age and composition. Monzogranite sheets were intruded into the Moine during the early stages of folding and metamorphism (Barr et al., 1985). Isotopic evidence from outside the district indicates that these events are probably Precambrian. The early metamorphism converted the monzogranite into strongly foliated gneissose granite. Subsequent reworking during the Ordovician stage of the Caledonian Orogeny resulted in the development of leucocratic segregations axial-planar to the later fold structures.

Metabasic rocks, postdating the gneissose granite, are common in parts of the Glenfinnan and Loch Eil groups (Rock et al., 1985). Discontinuous bodies of amphibolite are particularly abundant in the Achnaconeran Striped Formation. Most lie parallel to the bedding of the metasedimentary rocks and were probably intruded as sills. They were disrupted by subsequent deformation.

Xenoliths of gneissose granite, metasedimentary rocks and metabasic lithologies are common within several intrusive bodies of quartzose amphibolite which crop out within the gneissose granite south-west of Fort Augustus (Highton, 1994). The intrusions have unusual compositions and mineral assemblages, with abnormal concentrations of allanite, sphene and zircon. The origin of these intrusions is problematical, but may be a manifestation of alkaline igneous activity at the onset of the Ordovician stage of the orogeny.

During the waning stages of the Caledonian Orogeny, a period of extensive calc-alkaline magmatic activity commenced. This extended from the middle Silurian to early Devonian times. The metamorphic Caledonides, both north-west and south-east of the Great Glen Fault Zone, were intruded by plutonic bodies, vein complexes and dykes of a diverse range of compositions. To the north-west of the fault the earliest manifestations of this intrusive activity is the emplacement of biotite-microgranite, pegmatite and leucogranite veins of the Glen Moriston Vein Complex. These are most abundant in the Glenmoriston area but extend throughout the southern part of the district. The veins crosscut the foliation in the Moine, but show evidence of deformation and recrystallisation. Rocks ranging from feldspar-phyric microgranodiorite to mafic microdiorite form sheets (dip < 60°) and dykes (dip > 60°) that cut the intrusions of the Glenmoriston Vein Complex. Many of these intrusions have an igneous texture; however, most show evidence of deformation and partial recrystallisation. The more mafic members of the suite tend to predate the more felsic ones. Appinitic diorite intrusions are coarse grained, but have mineralogical and textural affinities with the mafic microdiorites. Several vents infilled with breccia occur in the southern part of the district. The clasts which make up the breccia are mostly of local origin, spalled from rock forming the wall of each vent. There is little evidence of extensive vertical transportation in these vents. All lie within the limit of the Glen Garry Vein Complex, and are thought to be of similar age. Large (> 100 m wide) masses, as well as a network of branching veins, form the Glen Garry Vein Complex. Most of the veins are granodioritic but the complex as a whole varies from diorite to monzogranite. Geophysical evidence suggests that the vein complex may, in part, represent the roof zone to a plutonic body. Emplacement was penecontemporaneous with a compositionally similar suite of microdioritic intrusions.

Most of the igneous rocks to the south-east of the Great Glen Fault Zone form part of, or relate to, the Foyers Plutonic Complex, mainly exposed in the Foyers district (Sheet 73E). The main component of this plutonic body in the Invermoriston district, is a hornblende-biotite granodiorite that forms both large irregular masses and ramifying veins. A large stock-like intrusion of appinitic diorite is also present. Minor intrusions are rare and represented by a few dykes and sheets of felsite and microdiorite.

Devonian sedimentary rocks are represented by the Old Red Sandstone, mainly in the Mealfuarvonie area.

The breccio-conglomerates were deposited in alluvial fans formed along fault-scarps bounding a narrow NE-trending basin. Finer-grained sediments were laid down beyond the toes of the fans. Near Drumnadrochit, within the Foyers district (Sheet 73E), these have yielded a rich miospore assemblage of late Lower Devonian to earliest Middle Devonian (Emsian to earliest Eifelian) age. Conglomerate, breccia and arkosic sandstone also occur in a fault-bounded strip south-west of Fort Augustus.

The Great Glen Fault Zone operated as a strike-slip structure during the later stages of the Caledonian Orogeny. It was reactivated, not only during the Devonian but also during later periods. The net displacement (sinistral) may exceed 100 km (Johnstone and Mykura, 1989). Within the fault zone, the Moine rocks were locally transformed into cohesive cataclasite during pre-Devonian movement. Syn- and post-Devonian displacements led to a general shattering, and the local development of fault breccia and clay gouge. The Strathglass Fault is also a major strike-slip structure with associated fracturing and reddening of the adjoining Moine rocks. The amount of displacement on this structure is uncertain.

A few camptonite dykes of late-Carboniferous to Permian age cut the Moine and the late-Caledonian igneous intrusions in the south-west part of the district. Their east–west trend is similar to that of dykes of the same age in other parts of the Northern Highlands.

The effects of Quaternary glaciations on the land surface are profound. Many of the landforms and deposits probably date from the maximum of the Main Late Devensian glaciation some 18 000 before present (BP). Ice movement was towards the north-east in the Invermoriston district where even the highest mountains show evidence of ice action. The district was completely ice-free for a period before valley glaciers re-advanced along Glen Moriston and the Great Glen from an ice sheet lying to the west and south-west during the Loch Lomond Stadial. Extensive glaciofluvial deposits were laid down during the final retreat.

TWO

Moine Supergroup

The Moine Supergroup, deposited originally as a succession of clastic sediments on a basement of Precambrian gneisses (Lewisian), was deformed and metamorphosed during the late Proterozoic and subsequently reworked during the Lower Palaeozoic stage of the Caledonian Orogeny. The Lewisian basement was incorporated within the Moine as infolds and tectonic slices and is exposed in parts of the Northern Highlands, but it does not crop out in the Invermoriston district.

LITHOLOGY AND STRATIGRAPHY

The Moine comprises psammite (quartz-feldspar granofels) with subordinate semipelite and pelite (mica schists or commonly gneisses) and also rhythmically interbedded psammite and pelite (striped schist). Calcareous rocks are absent apart from nodules of calc-silicate rock, probably metamorphosed concretions, which are important because of their value as indicators of metamorphic grade. Stratigraphical analysis is difficult because of the monotonous nature of the lithologies and the lack of distinctive units to aid correlation and mapping in areas of structural complexity.

A tectonostratigraphical framework by Johnstone et al. (1969) subdivided the Moine into the Morar, Glenfinnan and Loch Eil divisions. These are now regarded as formal lithostratigraphical groups (Roberts et al., 1987; Holdsworth et al., 1994). The Loch Eil and Glenfinnan groups, linked by a sedimentary transition zone, are represented in the Invermoriston district. Both groups share a common structural, metamorphic and igneous history.

Glenfinnan Group

A sequence of variably gneissose psammite, gneissose pelite and interlayered gneisses in the north-west corner of the district (Figure 3) lies within a belt of steeply inclined rocks which can be traced south-westwards to the type area in Glenfinnan (Strachan et al., 1988). The Glenfinnan Group in Glen Affric and Glen Cannich, including the parts lying within the Invermoriston district, were surveyed in considerable detail by Tobisch, (1963). His basis for classification was the bulk composition of a rock, taking little account of layering. Consequently, units separated as interbedded lithologies in the Invermoriston district were all classified as semipelite (Tobisch, 1966; 1967). The lithological units are tightly folded; because their stratigraphical position within the group is uncertain, they have not been assigned to formations. The intensity of deformation varies with rock type, the more pelitic units showing much tight folding and a high degree of strain with tectonic thickening in

fold hinge zones. The psammites are also tightly folded on a large scale, but internally show less strain. Cross-bed foreset angles are commonly preserved, and average 18° on fold limbs. They may be steepened by rotational strain in fold hinge areas to around 90°. Ductile shearing has occurred at major lithological junctions as shown by stratigraphical discontinuities, localised development of migmatite and attenuation of the lithological layering.

Notable exposures of the Glenfinnan Group rocks occur in parts of Coire Dubh [28 31], where cross-bedding is well displayed in relatively clean pavements continuing through several hundreds of metres of an apparently unbroken psammitic succession.

The rocks of the Glenfinnan Group in the north-west of the district are overlain by the Loch Eil Group. The latter is disposed in a major synform and the underlying Achnaconeran Striped Formation in the east limb of this structure is consequently assigned to the Glenfinnan Group (Figure 3).

ACHNACONERAN STRIPED FORMATION

The Achnaconeran Striped Formation comprises a thick sequence of rhythmically interlayered psammite and semipelite in very variable proportions, e.g. at Invermoriston [419 166]. Calc-silicate rocks are absent. The formation is divisible into facies, based on the relative proportions of psammite and semipelite, which grade imperceptibly into each other. It is difficult to estimate the relative proportions of each facies but the general overall impression is that psammite is more abundant than semipelite.

Facies 1 is made up of more than 60 per cent psammite and sequences up to 100 m thick of almost 100 per cent psammite, in beds up to 0.5 m thick, are common. The psammite is fairly coarse grained, possibly an original feature in part, although easily recognisable detrital grains do not occur.

In **facies 2** the relative proportion of psammite and semipelite varies from approximately 60:40 to 40:60. It is characteristically interlayered on a scale of approximately 1 to 5 cm and this structure is extremely prominent in fresh and weathered exposures, the psammite standing out in relief (see frontispiece). Each layer is normally bounded by parallel surfaces and can be traced along strike for 5 m or more without variation in thickness. The psammitic layers locally show cross-lamination and asymmetrical ripples on their upper surfaces (Figure 4). The psammite is commonly garnetiferous. Sparse muscovite porphyroblasts are enclosed by a pale coloured halo. The semipelitic layers are schistose to weakly gneissose and normally, but not invariably, garnetiferous. They also contain muscovite porphyroblasts, 1 to 2 cm across, which are particularly conspicuous in the western part of the outcrop of the formation where they crowd the foliation sur-

Figure 3 Map and cross-section showing the stratigraphy of the Moine.

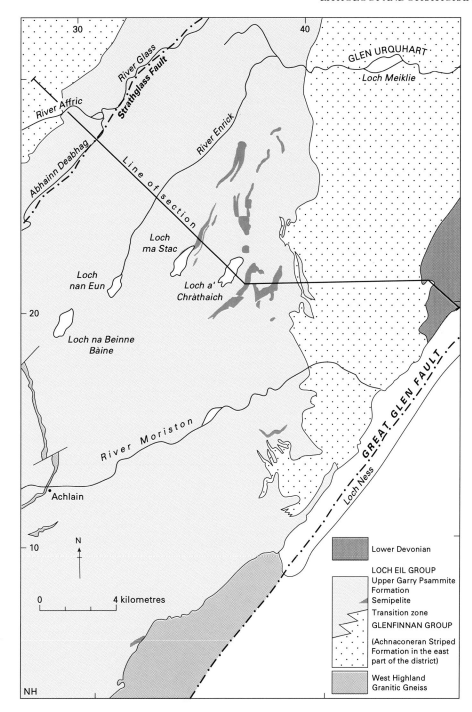

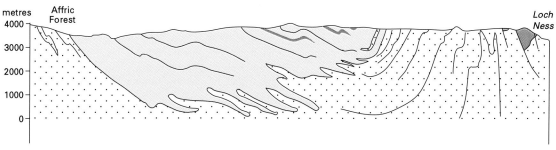

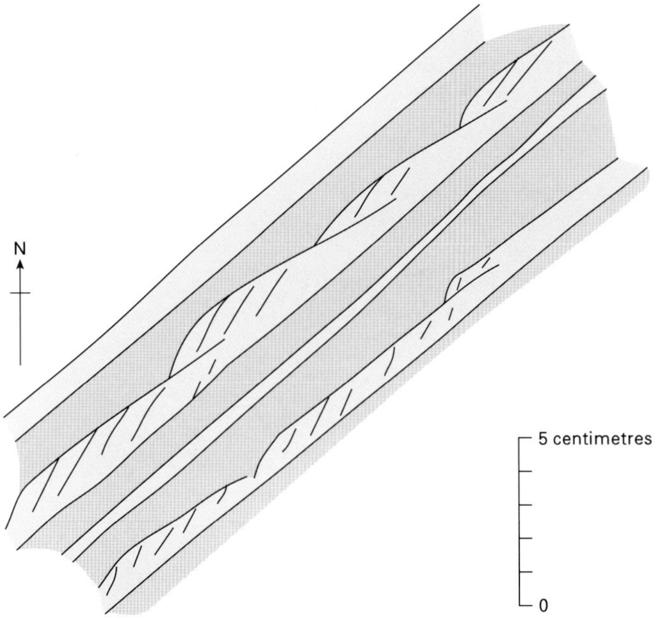

N

5 centimetres

0

Figure 4 Sedimentary structures in the Achnaconeran Striped Formation. Interlayered psammite and semipelite, the former showing cross-lamination and rippled surfaces draped by semipelite, north-east of Achnaconeran [4201 1860].

faces. The porphyroblasts have been modified tectonically into augen, well seen on surfaces normal to the foliation.

Semipelite constitutes more than 60 per cent of **facies 3**, and sequences of fairly homogeneous gneissose semipelite more than 100 m thick are present. However, a few widely spaced layers of micaceous psammite can usually be found. The semipelite is typically garnetiferous and crowded with muscovite porphyroblasts, modified as described above, particularly in the east. No original sedimentary textures or structures are preserved. Pegmatite pods and veins are present and become more abundant towards the east.

The exposed thickness of the Achnaconeran Formation is at least 3 km. The junction with the over-lying Loch Eil Group (Upper Garry Psammite Formation) is transit-

ional, over a thickness of 1.5 km. The transition zone shows large- and small-scale interbedding of striped rocks (facies 2) and psammite with the characteristic features, notably the presence of calc-silicate nodules, of the Upper Garry Psammite Formation. Sedimentary structures (where present) within the transition zone consistently young towards the Loch Eil Group.

Loch Eil Group

UPPER GARRY PSAMMITE FORMATION

The Loch Eil Group within the Invermoriston district comprises the Upper Garry Psammite Formation (Roberts et al., 1987). Its base in the north-west part of the district is taken at the contact between psammite and a thick band of pelite assigned to the Glenfinnan Group. The type lithology is a fine- to medium-grained psammite with thin semipelitic interbeds. Beds of psammite, commonly 2 to 30 cm thick, alternate with somewhat thinner layers of micaceous psammite, the latter more readily weathered than the former. In quarried exposures the psammite tends to break into slabs along highly micaceous partings (Plate 1). More massive, vaguely bedded psammite also occurs. Some of the beds are cross-laminated in tabular sets 1 to 10 cm thick but most are structureless or parallel laminated. Cross-lamination is seen only in well-washed exposures, particularly good examples being displayed at times of low water on the shore of Loch a' Chràthaich. Broad, shallow depressions in the tops of some of the beds are probably the result of penecontemporaneous erosion. A 10 cm bed of intraformational breccia has been recorded at Loch ma Stac [3376 2148] but this appears to be a solitary occurrence. Original sedimentary grains in the psammite have been completely recrystallised, the resulting rock being uniformly fine grained with evenly disseminated biotite in the more micaceous beds. Calc-silicate pods, up to

Plate 1 Finely interlayered psammite and micaceous psammite, with thin weakly segregated beds of semipelite, Upper Garry Psammite Formation. The thin, boudinaged sheets of schistose amphibolite locally cross-cut bedding. Road cutting on A887 [270 124]. (D 5068)

several centimetres thick and elongated parallel to the bedding, are invariably present. They commonly show mineral zoning with a garnet- and biotite-rich central part surrounded by a pale coloured rim. Occasionally they are seen to be discordant with respect to cross-lamination, providing evidence that they originated as calcareous concretions during diagenesis.

Psammite, with exceptionally abundant beds of micaceous psammite and semipelite, forms a complexly folded unit on Càrn Bingally [34 29].

Three semipelitic units, each up to 100 m thick, and several thinner units, crop out within the Upper Garry Psammite around Loch a' Chràthaich (Figure 5). They show a number of highly distinctive sedimentary and metamorphic features. The most notable primary characteristic is a millimetre-scale colour lamination, quartzose laminae alternating with darker, more continuous, biotitic laminae. The rock is fine grained, a texture inherited from the original sediment (silt and fine sand). The most prominent metamorphic mineral, kyanite, forms granular aggregates commonly up to 50 mm long and 15 mm across, and exceptionally more than 100 mm in length. The kyanite stands out in relief on weathered surfaces but is more difficult to locate in freshly broken rock. Garnet is also abundant. Spots up to 5 mm in length, crudely rectangular in outline with cores of muscovite and quartzose rims, are also a very characteristic metamorphic feature but of uncertain origin. Variations in the numbers of kyanite aggregates and muscovitic spots give rise to an ill-defined layering reflecting compositional variation in the original sediment. Calc-silicate rock is absent except for rare pods close to the junctions with the psammite above and below.

Cross-sections drawn across the Glen Moriston area (Figure 3) demonstrate an extremely rapid thinning of the Loch Eil Group in an easterly direction. One possible explanation of this arrangement is that the lower parts of the Upper Garry Psammite are laterally equivalent to the upper parts of the Achnaconeran Striped Formation.

The thick, dominantly sandy sequence of the Loch Eil Group may have accumulated under either fluviodeltaic or shallow-marine conditions, or a combination of both these environments. The existence of herring-bone cross-bedding (Strachan et al., 1988) indicates that parts of the succession were deposited in a regime subject to tidal activity. The uniformly small-scale nature of individual sedimentation units implies deposition of many hundreds of metres of sediment by low-velocity currents; this is consistent with accumulation in a shallow-marine environment. The kyanite-bearing semipelite can also be integrated into a shallow-marine shelf model by suggesting that it represents areas of the shelf characterised by high rates of silt deposition and starvation of sand (Strachan et al., 1988).

Moine rocks at Balnacarn

The gneissose granite sheet near Balnacarn [273 131] rests on coarse-grained psammite with persistent micaceous laminae which impart a regular flaggy structure to the rock. The psammite contains quartzofeldspathic leucosomes particularly in the micaceous layers. Calc-

silicate rocks are absent. Coarse-grained psammite also occurs above the gneissose granite sheet at Torr an Eas [285 145]; it forms an outcrop up to 350 m wide bounded to the south by a fault and to the east by a zone of shattering associated with a major structure, probably a thrust. The stratigraphical position of the coarse-grained psammite at Balnacarn is uncertain; it is probably migmatised Upper Garry Psammite Formation but the lack of calc-silicate rock suggests that it may be a separate formation.

STRUCTURE

Three phases of folding (D_1–D_3) can be recognised within the Moine of the Invermoriston district. Their ages are not precisely defined; D_1 and D_2 are probably Precambrian in age and D_3 Caledonian (Ordovician). The main structure is a major D_2 synform, with the Loch Eil Group in its core (Figure 3) and the Achnaconeran Formation (Glenfinnan Group) forming the eastern limb. The distribution of the lithological units within the Glenfinnan Group to the north-west of the Strathglass Fault is controlled by two major D_2 folds which Tobisch (1966) described as reclined (sideways closing) structures.

D_1 structures

Minor D_1 isoclinal folds occur in the Achnaconeran Formation, but are uncommon, and larger folds of this generation appear to be absent. Isolated intrafolial folds occur within the Upper Garry Psammite Formation. In some cases these are probably synsedimentary structures, although, where it can be shown that several beds are affected by a single fold, a tectonic origin is probable. It is rarely possible to measure the orientation of the fold axes and hence there is very little data concerning fold plunge and vergence.

A well-developed schistosity lying parallel to the bedding planes and crenulated in the cores of D_2 folds is the main evidence of D_1 deformation. The Upper Garry Psammite is characteristically flaggy due to the accentuation of the original bedding by micaceous laminae with aligned flakes of biotite and muscovite. In D_1 fold closures the mica-fabric is axial planar and cuts across the bedding.

The gneissose granites possess a D_1 penetrative fabric, and subconcordant quartzofeldspathic leucosomes are thought to represent syn-D_1 segregations within an initially homogeneous protolith (Barr et al., 1985). Schistose amphibolite sheets, subconcordant with the S_0/S_1 foliation in the adjoining metasedimentary rocks or the D_1 fabric in the gneissose granite, preserve a planar fabric. This is defined by amphibole and biotite crystals, aligned parallel to the D_1 fabric in the adjacent country rock (Figure 7).

A penetrative mineral lineation is well seen in many of the calc-silicate pods of the Upper Garry Psammite but in other rock types it is poorly developed. Although it has a similar orientation to the axes of D_2 folds (see below), the latter are characterised by axial crenulations of the D_1 schistosity and the lineation is probably a D_1 structure. Rodding is locally present in the Achnaconeran Striped Formation, for example in Allt Loch an t-Sionnaich [432 200]. Here the rodding lies approximately parallel to the axes of D_2 folds, a relationship which is considered to be fortuitous because at other localities, e.g. [4037 2155], rodding has been folded during D_2.

D_2 structures

The disposition of the Glenfinnan and Loch Eil groups within the Invermoriston district is controlled by D_2 folds. In the Affric Forest area Tobisch (1966) recognised two major tight to isoclinal folds. He showed that they are D_2 reclined structures responsible for the repetition of a pelitic unit. Elsewhere, including the Invermoriston district, younging directions, inferred from cross-bedding, indicate that there are no large-scale reversals due to earlier isoclinal folding and, apart from the local overturning of the limbs of intermediate-scale D_2 folds, the strata are right-way-up. On the largest scale the structure is a simple asymmetrical synform (Figure 3) with a steeply dipping eastern limb, a gently dipping western limb and a north-trending axial plane trace approximately in the position of grid line 38. Intermediate-scale folds in the axial zone of the main structure have been revealed by the mapping of semipelitic units (Figure 5). Minor folds are unevenly distributed. In the eastern limb of the synform they are abundant, although there are zones of uniform steep dip. On the western limb, there are broad belts in which folds are completely absent and the rocks maintain a uniform dip of 20 to 30° to the south-east. The attitude of the fold axes and axial surfaces together with the style and vergence of the minor folds varies systematically across the district (Figure 6). Individual minor folds show pronounced variations in style along their axial plane trace, from tight to open structures. To the north of Glenmoriston there is also a west-to-east contrast in style, with fairly open folds in the Upper Garry Psammite Formation giving way to tight to isoclinal folds in the Achnaconeran Striped Formation. Elsewhere the D_2 folds are consistently tight. The non-cylindrical nature of the D_2 folds has led to the development of curvilinear fold axes which manifests itself in variations in the amount and direction of plunge (Plate 2), both at outcrop level

Figure 5 Stratigraphy and structure of the Moine near Loch a' Chràthaich.

(Figure 5, semipelitic unit B) and individual folds within exposures. In the isoclinal folds within the Achnaconeran Striped Formation the variation of plunge is very large. The curvilinear axes cannot be interpreted as the result of later refolding nor the superimposition of D_2 folds on earlier structures but are considered to be a primary characteristic of the D_2 deformation.

A weakly to well-developed axial planar structure is displayed in folded beds of psammite on the south face of Binnilidh Bheag [326 154] and just east of Cnoc Liath [3088 1437]. In some of the micaceous layers the D_2 structure is a crenulation of the D_1 schistosity; in others the early schistosity is completely transposed to produce a new continuous fabric (S_2). An intense crenulation and transposition of the D_1 fabric also occurs in the semipelitic layers of the Achnaconeran Striped Formation.

The relationship of the D_1 fabric to the D_2 folds can be seen in the gneissose granite near Torr an Eas [2809 1400]. Here the D_1 fabric and concordant sheets of schistose amphibolite are folded (Figure 7). The gneissose granite has developed a new schistosity in thickened hinge zones and the D_1 fabric of the schistose amphibolite is crenulated. Similar relationships are displayed in the Fort Augustus gneissose granite at Lon Mór [33 79] where new quartzofeldspathic segregations have developed axial planar to D_2 folds.

A mineral lineation is commonly seen in psammite as well as calc-silicate pods in the Loch Eil Group. In places,

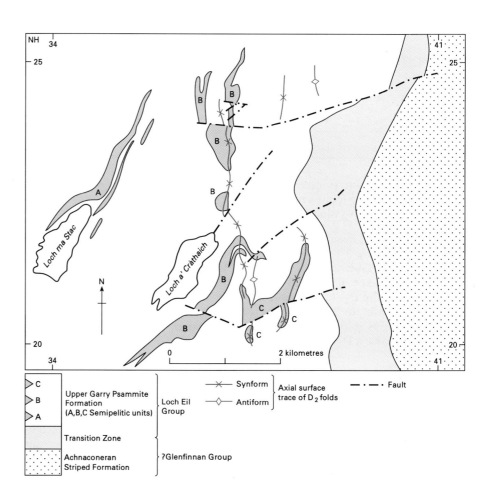

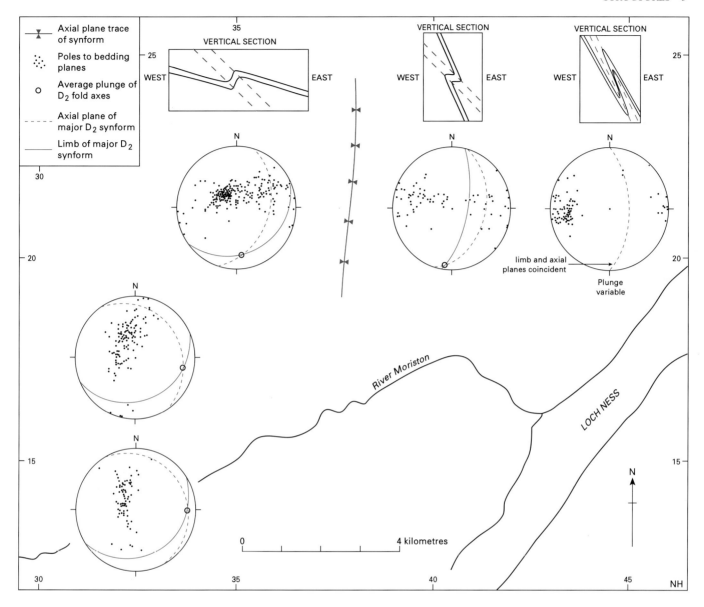

Figure 6 Diagrammatic cross-sections and equal area, lower hemisphere projections showing the orientation of D_2 structures north of Glen Moriston. The orientation of the axial plane is based on data derived from D_2 minor folds.

e.g. Binnilidh Bheag [32 15], it lies approximately parallel to the axes of D_2 folds and, on this basis, could be assigned to the D_2 deformation. However, in sheets of schistose amphibolite the lineation is a component of a fabric which was folded and crenulated during D_2 and this appears to be a D_1 structure.

D_3 structures

In highly micaceous lithologies crenulations, which are discordant with respect to D_2 folds, are presumed to be D_3 structures. North-west-trending D_3 folds, the Monar Phase of Tobisch et al. (1970), affect the Glenfinnan Group rocks of Glen Affric. These folds die out rapidly towards the south-east and the Upper Garry Psammite

and Achnaconeran Formations lie in an area of low D_3 strain. The boundary between these structurally contrasting areas marks the eastern limit of severe (probably Ordovician) tectonic reworking of early (probably Precambrian) structures (the Loch Quoich Line of Roberts and Harris, 1983; Roberts et al., 1987).

METAMORPHISM

The timing and metamorphic history of the Northern Highland Moine during the Caledonian Orogeny are poorly constrained (cf. Winchester, 1974; Fettes et al., 1985). There is a general absence or poor preservation of metamorphic index minerals in pelitic and semipelitic

Plate 2 Curvilinear D$_2$ fold pair in rocks of the Upper Garry Psammite Formation. A887 road cutting near Achlain [276 124]. (D 5071)

lithologies. This largely reflects the paucity of lithologies with compositions suitable to develop index minerals (Barr, 1983). Further, a widespread retrogression of these early metamorphic assemblages occurred during the syn-to post-D$_2$ period. Most studies of the metamorphic history of the Moine have been concerned with the distribution of index minerals in calc-silicate assemblages (Kennedy, 1949; Soper and Brown, 1971; Winchester, 1974; Tanner 1976). Powell et al. (1981), in studies from the Morar/Glenfinnan area, presented a fourfold division of the calc-silicate assemblages, with prograde and retrograde assemblages distinguishable on textural grounds. To the east of the Loch Quoich Line (Roberts and Harris, 1983), the metamorphic history of the Sgurr Beag Nappe is equivocal. Winchester (1974), noted the presence of some middle amphibolite facies, kyanite-bearing assemblages within the Loch Eil Group. However, most of its outcrop, including the Invermoriston district, was thought to comprise of lower grade, garnet-zone rocks.

Petrography of metamorphic index mineral-bearing lithologies

CALC-SILICATE ROCKS

The occurrence of calc-silicate-bearing rocks is restricted to psammitic or micaceous psammitic lithologies of the Upper Garry Psammite Formation. They typically form coarse-grained, cream or pale grey pods, lenses or layers up to 200 mm in thickness. Most contain red or cinnamon-coloured garnets up to 5 mm in diameter, and elongate spindles and aggregates of mafic minerals (most commonly amphibole) up to 6 mm in length. These are set in a white or pink-white granular matrix of quartz and plagioclase feldspar. A crude concentration

Figure 7 North–south vertical section showing D$_2$ folding of gneissose granite and schistose amphibolite, near Torr an Eas [2809 1400].

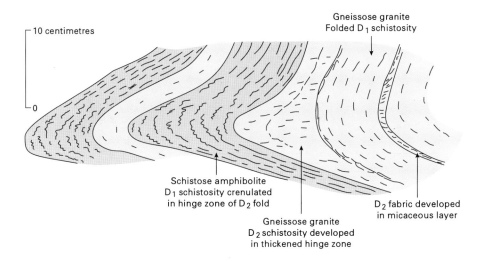

of garnet defines a compositional layering in the thicker beds and lenses, e.g. in the Allt Dail à Chùirn [3059 0608] (S 71388). Elongate mafic minerals, and stretched garnets (S 71318, 71319), define a linear fabric parallel to the D_2 fabric in the host rocks, e.g. at [2928 0471]. Most contacts with the host rocks are sharp, although at [3350 0728 and 3415 0707] calc-silicate pods grade over several centimetres into the host psammite, overgrowing the S_0–S_1 foliation.

Calc-silicate assemblages similar to those described by Powell et al. (1981) are found. The following parageneses are recognised:

(1)   Biotite + garnet + plagioclase + quartz
(2a)   Hornblende + garnet + plagioclase + quartz
       ± pyroxene + zoisite + biotite
(2b)   Hornblende + garnet + plagioclase + quartz
       ± pyroxene ± actinolite ± clinozoisite
       ± zoisite

Sphene, apatite and opaque minerals are ubiquitous with minor ilmenite, zircon and pyrite. Type 2a assemblages, containing relict clinopyroxene and/or plagioclase with compositions more calcic than An_{70}, occur in a few rocks from the northern part of the district (Figure 8a, b). This suggests that the Moine succession here was subjected to an early, syn-D_1 to pre-D_2, high-grade metamorphic event. Over much of the district both type 1 and 2b assemblages are predominant, with feldspar compositions largely in the range An_{30} to An_{50}. Textural evidence suggests these are disequilibrium assemblages, and in part retrogressive. Where preserved, clinopyroxene forms either xenoblastic aggregates or small grains. In most assemblages a mid-green to green-brown to pale green hornblendic amphibole replaces the pyroxene. This amphibole is the predominant mafic mineral in most calc-silicate rocks, forming either large poikiloblastic grains sieved with quartz, or elongate aggregates of idioblastic crystals. Poikiloblastic garnets are ubiquitous. Most porphyroblasts are two-stage, with syn-D_1 idioblastic, inclusion-rich cores, overgrown by largely inclusion-free, syn-D_3 rims (S 72142). Fine-grained inclusions of quartz, plagioclase, opaque minerals, minor amphibole and/or biotite within the garnet cores often have a grain shape fabric that picks out discrete inclusion trails. Single-stage, allotriomorphic garnet porphyroblasts, which overprint the D_2 fabric, are common. Their presence suggests that further growth of garnet occurred during D_3. Zoisite is widespread (Figure 8c), forming both inclusion-rich, prismatic porphyroblasts and quartz-zoisite symplectites after plagioclase. The age of zoisite growth is equivocal, with the porphyroblasts having no preferred orientation, and no discernible alignment of inclusions.

In most rocks the early-formed assemblages are metastable with retrogression common, although the timing of this event is uncertain. The hornblendic amphibole is mostly stable; however, the crystals are locally overgrown or replaced by irregular aggregates of actinolite and/or biotite, with sphene and more rarely clinozoisite (S 71318). Rare biotite pseudomorphs, after amphibole

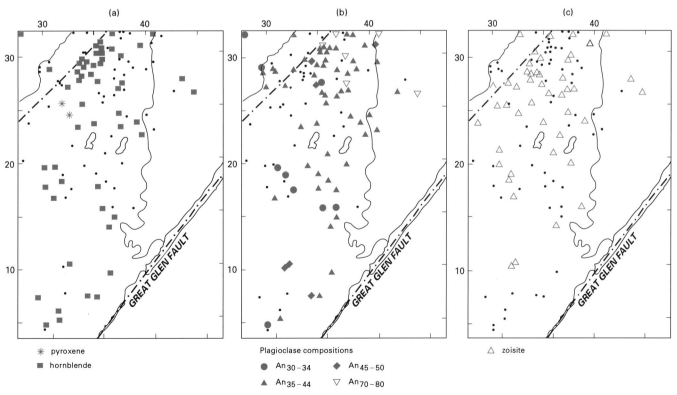

Figure 8   Distribution of index minerals in Moine calc-silicate lithologies.

(S 72082), take the form of either single, large bladed crystals or aggregates up to 3 mm, elongate parallel to the D_2 lineation. Clinozoisite is present as irregular grains or aggregates, mostly in plagioclase altered to sericitic mica. The clinozoisite is mainly a product of feldspar retrogression (cf. Powell et al., 1981). In some rocks, though, it epitaxially overgrows small cores of allanite (S 72082). White mica, chlorite and calcite are secondary minerals in these rocks.

SEMIPELITES

Indicators of metamorphic grade within semipelitic rocks of the district are sparse, with rare kyanite and/or staurolite-bearing assemblages (Figure 9). These assemblages represent the earliest (M_1), Precambrian, tectonothermal event in the district. Contemporaneous with this is the development of the regional D_1 foliation and gneissose segregations in the metasedimentary rocks and granitic gneiss. In most rocks replacement of the index minerals by white mica is generally extensive. Kyanite

(S 71361, 71391, 71323, 74839) and rare staurolite (S 71361) are generally present only as corroded, relict grains within syn- to post-D_2 muscovite porphyroblasts. However, the semipelites within the upper part of the Upper Garry Psammite Formation outcropping in the Loch a' Cràhaich area (Figure 5) contain well-preserved subidioblastic crystals or granular aggregates of kyanite (S 71505, 71508). The semipelitic assemblages contain evidence for a later metamorphic overprint (M_2). Fibrolitic sillimanite needles occur within muscovite plates (S 71509), and replace the muscovite partially enclosed within the outer rims of late garnets (described below). The timing of fibrolite growth is equivocal.

Garnet porphyroblasts are idioblastic to xenoblastic, mostly composite crystals, with two discrete growth zones. The early formed (Gt_1) garnets are subidioblastic, single-stage, mostly inclusion-free crystals. Where inclusions trails are present, they are oblique to the enclosing composite S_0/S_1 and S_1/S_2 schistosity (S 71404, 72097, 74845), suggesting a syn-D_1 to pre-D_2 age. These early formed garnets are metastable, with disequilibrium suggested by resorption. This takes the form of atollisation of some garnets, and replacement by biotite, quartz and plagioclase symplectites (S 72098). All have reverse zoning profiles of decreasing Mg and increasing Fe from core to rim (Figure 10), although this is less distinct in

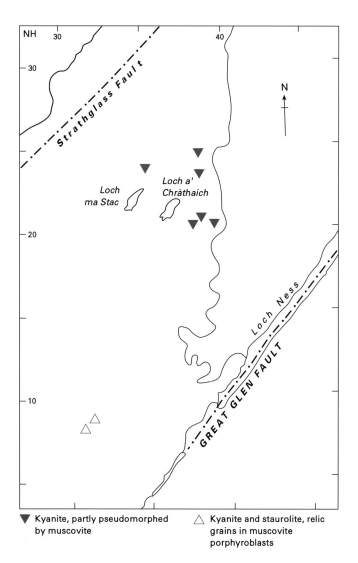

Figure 9 Distribution of metamorphic index minerals in semipelitic rocks in the Moine.

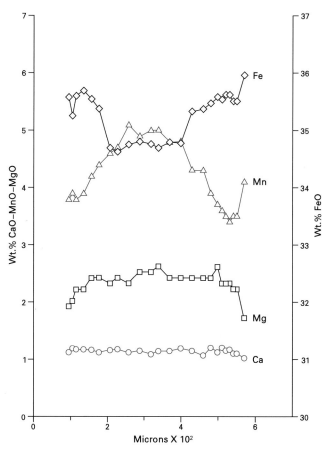

Figure 10 Electron microprobe traverse through an early formed garnet (Gt_1).

the smaller subidioblastic garnets. Where the semipelitic rocks are muscovite-poor, only single-stage Gt_1 garnets are found. In muscovite-rich rocks, the garnets vary in morphology from idioblastic to xenoblastic crystals elongate within the foliation, but predominantly occur as large (up to 5 mm) composite crystals. These comprise pale coloured Gt_1 cores, enclosed by dark, inclusion-rich rims (Gt_2). Many rims are synkinematic, with sygmoidal inclusion trails, although locally they overgrow the regional S_2 crenulation fabric, showing that in part they postdate D_2 deformation. These trails of inclusions wrap the Gt_1 cores, and are oblique to any early fabrics preserved. The garnets have a distinct compositional profile, with a discontinuity at the boundary between the two stages (Figure 11).

White mica is ubiquitous, and mostly replaces biotite. Towards the interface between the two micas, biotite compositions decrease in Fe, Mg and Ti, with concomitant increase in Si, Al, Na and K. The boundary is a transition, passing into a mixed 'biotitic' muscovite. Exsolution of Ti not held within the new mica structure takes the form of fine grains or laminae of ilmenite. These occur at both the interfaces between the micas, and along the cleavage planes of the new muscovites (S 71351, 71363). The growth of muscovite is syn- to post-

D_2 in age. There is evidence of mimetic replacement and/or overgrowth of biotite in both S_2 crenulation cleavage (S 70735, 72103) and composite S_1/S_2 (S 72099, 72137). The development of the white mica porphyroblasts is contemporaneous with muscovite growth in the matrix. In rocks containing index minerals, the white mica mostly forms aggregates of either microcrystalline white mica (S 71298) or small crystals of muscovite (S 71361). These enclose corroded grains of kyanite and/or staurolite (cf. Barrow, 1893; Neathery, 1965; Chinner, 1966; Yardley and Baltzais, 1985). The fine-grained micas coarsen either to aggregates of muscovite plates (S 71323, 71506) or muscovite books with single flakes up to 20 mm across (S 68548, 68551, 71323, 71506). The consuming reaction is rarely complete, with the index minerals preserved as small armoured relict grains in large single muscovite plates, which lie obliquely to the regional foliation. In other semipelitic lithologies, the muscovite porphyroblasts occur only within the quartzofeldspathic lits. Here, they form elongate, randomly orientated flakes enclosed by quartz-rich coronas. Both biotite and plagioclase are present only as corroded, relict grains in the quartz coronas (S 71351). A good example is found in the micaceous psammites exposed in the Allt Dail a' Chùirn [3045 0519], where muscovite-quartz ocelli up to 40 mm in diameter occur. White micas from the matrix, together with the porphyroblasts and those interleaved with biotite, are mostly phengite with a paragonite content of less that 7 mol.%.

The growth of fibrolite and, more rarely, acicular sillimanite (S 71361, 71509) is a later, possibly syn-D_3 metamorphic overprint, which mostly affects the muscovite porphyroblasts.

Metamorphic conditions

Estimates of temperature and pressure during the regional metamorphic events have been derived using published thermodynamic calibrations. These reactions are based on the activity models for the mineral parageneses commonly occurring in pelitic and semipelitic rocks (Hodges and Crowley, 1985; Gangully and Saxena, 1984). Errors of the order ±50°C and ±1 kb are inherent in the results due to uncertainty in the thermodynamic data and reacting systems. Apparent temperatures derived from assemblages bearing Gt_1 garnets are in the range 620–650°C. Results from single Gt_2 and composite Gt_1/Gt_2 garnet porphyroblasts are, however, slightly higher, in the range 670–690°C. Estimates of pressure are approximately 5 kb from Gt_1 assemblages, while for Gt_2 garnet-bearing rocks results cluster close to 7 kb.

Both calc-silicate and semipelitic rocks of the district, preserve mineral assemblages that, on textural grounds, appear to record two discrete lower to middle amphibolite facies events. M_1 metamorphism culminated in the growth of staurolite and kyanite in the peak metamorphic assemblage. It is contemporaneous with the D_1 deformational event and with localised gneissose segregations in the S_0/S_1 regional foliation of the metasedimentary rocks and Fort Augustus granitic gneiss. Both the absence

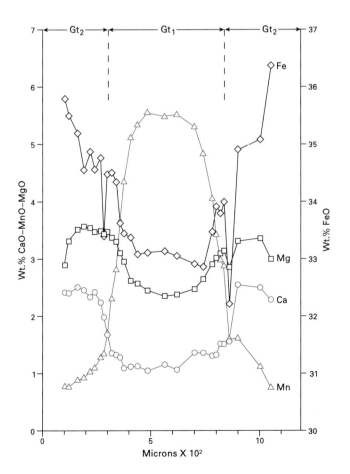

Figure 11 Electron microprobe traverse through a composite garnet (Gt_1–Gt_2).

of an Mg discontinuity in the Gt_1 garnets and the preservation of Mn-rich cores to the Gt_1 garnet porphyroblasts suggest a prograde metamorphic history.

Textural evidence suggests that the onset of the Gt_2 garnet growth is contemporaneous with the crystallisation of muscovite porphyroblasts. This reaction involves the breakdown of the index minerals staurolite and kyanite, with the partial replacement of biotite and calcic plagioclase. These reactions are in part degenerate, indicating disequilibrium (lower grade?) conditions either syn- to post-D_2. The M_1 index assemblages are largely recrystallised, with remnants preserved only as relic phases in the later M_2 assemblages. Thus, the derived estimates of pressure and temperature from the Gt_1 garnets are unlikely to reflect M_1 metamorphic conditions. The timing of Gt_2 growth is poorly constrained in these rocks.

The garnets in part postdate structures and fabrics associated with the D_2 deformational event, but are mostly syn-kinematic and contemporaneous with the growth of muscovite. The second metamorphic episode is poorly constrained, but probably contemporaneous with the D_3 (Ordovician) tectonothermal event to the west of the Loch Quoich Line (cf. Barr, 1983). The emplacement of the deformed pre-D_3 Glen Dessary Syenite at c. 456 Ma (van Breemen et al., 1979) constrains this event.

Widespread alteration of biotite and garnet to chlorite in the semipelites is attributed to post-D_3 retrogression. In the calc-silicate assemblages this retrogrograde metamorphism produces widespread saussuritisation of plagioclase, alteration of pyroxene and hornblendic amphibole to actinolite, and replacement of garnet, biotite and hornblende by chlorite.

THREE

Grampian Group

In the Invermoriston district, rocks of the Grampian Group only occur south-east of the Great Glen Fault Zone (Figure 12). It consists mainly of lower to middle amphibolite facies psammitic and semipelitic rocks. The lower contact of the group is defined by a ductile tectonic break, the Eilrig Shear Zone, which separates it from the underlying metasedimentary rocks of uncertain stratigraphical position (which are dealt with in Chapter 4). The upper boundary of the Grampian Group does not occur in the Invermoriston district, but is exposed to the south, in the Glen Roy district (Sheet 63W), where it is overlain by pelitic and quartzitic rocks of the Appin Group.

nae are interpreted as representing original bedding, virtually undisturbed by the development of the metamorphic mineral fabric. Cross-bedding and other sedimentary structures are locally common. Semipelite occurs as units up to several hundred metres thick. A schistosity is invariably well developed and original bedding can be recognised only rarely.

Rock units occur in which semipelite and psammite are present in almost equal proportions, the two lithologies being interbedded on a scale of 5–20 cm. As such, individual rock types cannot be separately mapped and formations of mixed lithologies have therefore been recognised.

LITHOLOGY AND STRATIGRAPHY

The Grampian Group of the Invermoriston district is made up of a conformable succession of psammites and semipelites with occasional thin beds of quartzite and pelite. Each of the main rock types forms a mappable unit that can be traced for several kilometres along strike. They also occur in interbedded assemblages. The units show rapid lateral facies changes and boundaries are nearly always gradational. Gradations from psammite to micaceous psammite to semipelite are recognisable.

The psammites are granular quartz-feldspar-biotite rocks with well-developed planar foliation. They are flaggy or massive, with parting planes developed along thin micaceous laminae, 1–2 mm thick, typically spaced at approximately 20 cm intervals. Biotite is disseminated throughout the rock but the content is variable, imparting a colour banding. The colour banding and the micaceous lami-

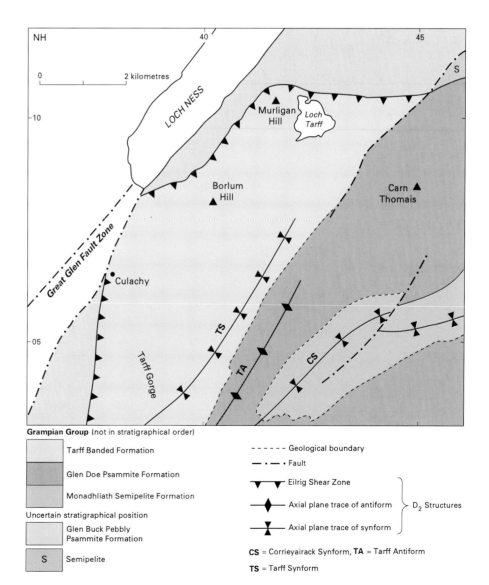

Figure 12 Grampian Group and rocks of uncertain stratigraphical position south-east of the Great Glen Fault Zone.

Grampian Group (not in stratigraphical order)

Tarff Banded Formation

Glen Doe Psammite Formation

Monadhliath Semipelite Formation

Uncertain stratigraphical position

Glen Buck Pebbly Psammite Formation

S Semipelite

- - - - - Geological boundary
- · - · - Fault
Eilrig Shear Zone
Axial plane trace of antiform
Axial plane trace of synform } D₂ Structures

CS = Corrieyairack Synform, **TA** = Tarff Antiform
TS = Tarff Synform

Calc-silicate layers in the psammitic and semipelitic rocks occur throughout the Grampian Group succession. They vary greatly in abundance and are lenticular, typically up to 50 mm thick, and taper out along-strike within 1 to 2 m. Some contain large porphyroblasts of green hornblende showing poorly defined dimensional orientation. The calc-silicate layers are interpreted as representing metamorphosed calcareous concretions.

A lithostratigraphy (Table 1) has been established for the Grampian Group in the Foyers district (Sheet 73E) to the east (Haselock et al., 1982), and in the Glen Roy district (Key et al., 1997). Continuity of outcrop with the successions in those districts has allowed the established lithostratigraphy to be extended to nearly all the Grampian Group rocks of the Invermoriston district, although the stratigraphical affinity of one unit, the Tarff Banded Formation, is uncertain.

GLEN DOE PSAMMITE FORMATION

The Glen Doe Psammite Formation is the lowest recognised stratigraphical unit and is made up of a monotonous sequence of psammitic rocks that occur in two outcrops separated by the stratigraphically younger Monadhliath Semipelite Formation. The formation is folded by major D_2 folds and the two outcrops occur in the limbs of the D_2 Corrieyairack Synform. There is some repetition of the rocks of the formation, which appears to be at least 2 km thick. The boundary with the Tarff Banded Formation is a normal lithological contact which, northwest of Càrn Thòmas [439 092], is disrupted by the Stratherrick–Loch Mhor Fault. The boundary with the Monadhliath Semipelite Formation is also a normal lithological boundary transitional over a thickness of some tens of metres.

The formation is made up of massively bedded psammite, with parting planes spaced at intervals varying from 60 mm to 1 m. The parting planes occur along thin layers of semipelite, up to 50 mm thick. The psammites are fine- to medium-grained quartzofeldspathic rocks with occasional beds of coarse material in which the quartz and feldspar grains reach up to 2 mm in diameter. Ripple cross-lamination in sets up to 60 mm has been recorded, and convolute bedding also recognised, but generally the psammites are massive, parallel-bedded, and without visible sedimentary structures.

Calc-silicate lenses occur within the psammites but they are uncommon. In the Loch Killin area of the Foyers district (Sheet 73E), Whittles (1981) divided the formation into an upper unit with white calc-silicate lenses and a lower unit distinguished by the presence of green calc-silicate layers. It has not been possible to extend this subdivision of the Glen Doe Formation into the Invermoriston district. In the top 250 m of the formation the psammites are more micaceous and flaggy, and separated by thicker more abundant semipelitic beds. Such interbedded lithologies are well exposed on Càrn Doire Chaorach [409 049], and to the south-west of the summit of Carn a' Chuilinn [417 034]. The interbedded sequences mark a transition to the more semipelitic lithologies of the Monadhliath Semipelite and Tarff Banded formations.

MONADHLIATH SEMIPELITE FORMATION

The rocks of this formation occur in the core of the Corrieyairack Synform where they form a single NE-trending outcrop bounded to the north-west, south and south-east by psammitic rocks of the Glen Doe Formation.

The formation is chiefly made up of schistose semipelite, interbedded with thin units of micaceous psammite. The semipelite is dark grey, well foliated, fine grained and consists of quartz, plagioclase, biotite, muscovite and garnet. The micas define the foliation which, except where folded by minor folds, is concordant with the bedding. The psammitic beds vary in abundance. Overall, psammite is less abundant than the semipelite and forms thin beds commonly 3–10 cm thick. Thin beds of quartzite also occur but rarely exceed 5 m in thickness and are impersistent along strike. Quartzite is an important component of the basal part of the formation, particularly in the hinge region of the Corrieyairack Synform. Calc-silicate lenses occur both in the semipelitic and psammitic lithologies.

TARFF BANDED FORMATION

The rocks of this formation are exposed in NNE-trending outcrop at least 3 km wide. The western boundary is defined by the Eilrig Shear Zone, separating it from the underlying Glen Buck Pebbly Psammite Formation. The eastern boundary is a normal lithological contact with the Glen Doe Psammite Formation. The

Table 1
Lithostratigraphy of the Grampian Group.

Glen Roy district (63W)*	Invermoriston district (73W)		Foyers district (73E)**
Auchivarie Psammite Formation			Carn Leac Psammite Formation
Tarff Banded Formation	Tarff Banded Formation	Monadhliath Semipelite Formation	Monadhliath Semipelite Formation
Glen Doe Psammite Formation	Glen Doe Psammite Formation		Glen Doe Psammite Formation
Coire nan Laogh Semipelite Formation			Coire nan Laogh Semipelite Formation

* Key et al. (1997)
** Haselock et al. (1982)

formation, probably more than 2 km thick, shows much repetition as a result of major D_2 folds. It consists of an interbedded sequence of semipelite and micaceous psammite, forming a mixed assemblage of units too thin and impersistent to be mapped separately. Subordinate beds of quartzite and pelite are present but rarely exceed 5 m in thickness and are impersistent along strike.

Overall the proportions of psammitic and semipelitic lithologies are approximately equal, although semipelite increases towards the north-east. Calc-silicate lenses, rarely more than 30 mm thick, are present throughout the formation. They are characterised by hornblende as large irregular porphyroblasts or as small blade-like crystals. Garnet porphyroblasts with snowball, spiral and 'S'-shaped inclusion trails occur in semipelitic rocks near the Eilrig Shear Zone. The inclusion trails are composed of fine-grained opaque minerals (graphite, ilmenite, magnetite) which are continuous with the mica-rich folia of the matrix.

LITHOSTRATIGRAPHICAL CORRELATION

The stratigraphical status of the Glen Doe Psammite and Monadhliath Semipelite formations is confirmed by continuity of outcrop with rocks of the type areas (Haselock et al., 1982). Younging evidence suggests that the Monadhliath Semipelite Formation is younger than the Glen Doe Psammite Formation. The Tarff Banded Formation has no continuity of outcrop with the established lithostratigraphy. Sedimentary structures near the contact with the Glen Doe Formation are rare and provide equivocal evidence for the age relationship between the two formations. The upper part of the Glen Doe Formation includes a sequence of interbedded psammite and semipelite, well exposed at the boundary with the Monadhliath Semipelite Formation at Càrn a' Chuilinn [414 035]. A similar sequence can be traced along the boundary with the Tarff Banded Formation, from [401 033] north-eastwards for some 3 km. Structural interpretations by Haselock et al. (1982) suggested that a major structure, the Tarff Antiform, repeats the lithostratigraphy. Thus, the lithological and structural evidence presented by these authors supports a correlation between the Tarff Banded Formation and the Monadhliath Semipelite Formation.

STRUCTURE

Within the Invermoriston district and the adjoining districts of Glen Roy and Foyers the distribution of the Grampian Group stratigraphical units is controlled by upright, open to tight major folds having NE-trending axial traces. These folds, formed during regional deformation. They fold bedding (S_0) and a bedding-parallel foliation formed during an earlier D_1 event. Most of these major folds have been described previously and assigned to D_2 locally (Anderson, 1956; Haselock et al., 1982; Key et al., 1997). Petrographical evidence indicates a regional increase in pressure-temperature conditions

during the D_2 event. The metamorphism peaked during amphibolite facies metamorphism and may have extended into post-D_2 times.

D_1 structures

The ubiquitous foliation throughout the district is regarded as a bedding-parallel D_1 planar fabric. This fabric is axial planar to rare minor isoclinal folds, but evidence for the existence of D_1 folding is sparse. Haselock et al. (1982) postulated the existence of D_1 fold closures in the Tarff Gorge. This conclusion was based on apparent changes of vergence of minor folds but has not been confirmed during the present survey.

D_2 structures

The distribution of the lithological units is chiefly controlled by major D_2 folds with NE-trending and steeply dipping axial planes. Minor D_2 folds are locally common and are typically tight, asymmetrical folds of variable plunge. Studies of the fold vergence has assisted the identification of associated major folds.

CORRIEYAIRACK SYNFORM

This structure was described by Anderson (1956) and, in greater detail, by Haselock et al. (1982). It has a NE-trending axial trace and controls the outcrop of the Monadhliath Semipelite Formation which occurs in its core, flanked by psammitic rocks of the older Glen Doe Psammite Formation (Figure 12). Associated minor folds plunge at low to moderate angles to the north-east and the fold hinge consists of several synforms and complementary antiforms.

TARFF ANTIFORM

This major fold was first described by Haselock et al. (1982) as the complementary antiform west of the Corrieyairack Synform (Figure 12). The fold is an asymmetrical structure, with a subvertical, locally overturned, north-west limb. Its axis is located within the outcrop of the Glen Doe Formation. The axial trace has been identified a short distance east of the boundary of the Tarff Banded Formation by vergence of minor folds.

TARFF SYNFORM

The Tarff Synform lies to the west of the Tarff Antiform and is located within rocks of the Tarff Banded Formation. It has a NNE-trending, subvertical axial plane. The related minor folds plunge at low to moderate angles to the NNE.

Eilrig Shear Zone

The Eilrig Shear Zone is a major ductile tectonic discontinuity that separates the Grampian Group from the Glen Buck Pebbly Psammite Formation. The shear zone consists of a variable sequence of mylonites and phyllonites interleaved with less-deformed psammitic rocks. It has been traced from Càrn an t-Suidhe [447 105] to the southern margin of the sheet, and thereafter through the Glen Roy district as far as the eastern slopes of the Great Glen above Loch Lochy, a distance of at least 20 km. It continues north-eastwards into the Foyers district (Sheet 73E).

The Eilrig Shear Zone is made up, for the most part, of intensely deformed mylonitised units of the underlying Glen Buck Formation. Locally, discontinuous layers of phyllonitic rocks are present, interpreted as mylonitised units of the Grampian Group incorporated into the shear zone.

The Eilrig Shear Zone is well exposed on the western slope of Borlum Hill [395 085] where a SE-dipping sequence, about 400–500 m thick, of protomylonitic pebbly psammites, sheared psammites and metaconglomerates, quartz-ribbon mylonites and quartz-feldspar mylonites comprise an almost continuous section through the shear zone. The increase in intensity of deformation upwards through the sequence is well displayed. Similar well-exposed sequences occur at Murligan Hill [413 107]. Rotation of garnet porphyroblasts in the Tarff Banded Formation, lying immediately above the shear zone, indicates overthrusting of the amphibolite facies Grampian Group north-westwards over the lower metamorphic grade rocks of the Glen Buck Pebbly Psammite Formation (Phillips et al., 1993).

METAMORPHISM

Phillips et al. (1993) report estimates of T = 410–575°C, P = 6.5–7.8 kb for garnet-bearing phyllonites and 525–620°C, P = 6.2–8.2 kb for mylonitic rocks exposed within the upper part of the Eilrig Shear Zone. These are broadly comparable with previously published estimates for amphibolite facies peak regional metamorphism for Appin Group Dalradian rocks from the Spean Bridge area to the south of the district, e.g. T = 535°C, P = 5 kb (Richardson and Powell, 1976); T = 540 ± 30°C, P = 6.5 ± 0.5 kb (Powell and Evans, 1983); T = 525°C, P = 5.0 kb (Wells, 1979). This metamorphism accompanied the D_2 deformation of the Grampian and Appin groups in the Spean Bridge–Glen Roy area (Key et al., 1991; Phillips and Key, 1992).

FOUR

Rocks of uncertain stratigraphical position

To the south-east of the Great Glen Fault, the Grampian Group succession is underlain by metasedimentary rocks of uncertain stratigraphical affinity, the Glen Buck Pebbly Psammite Formation, from which it is separated by the Eilrig Shear Zone. Other metasedimentary rocks of uncertain stratigraphical status are found as tectonically bound slices within the Great Glen Fault Zone.

GLEN BUCK PEBBLY PSAMMITE FORMATION

Coarse clastic arkosic psammitic rocks were first described within the Invermoriston district by Mould (1946). Green, epidote-rich schistose grits were noted on Beinn a' Bhacaidh [432 119] and grits interbedded with mica schists on Borlum Hill [396 085]. These are part of a sequence of well-bedded, coarse-grained, schistose gritty psammites, with interbedded semipelitic and pelitic lithologies. These were referred to by Parson (1982) as the 'Glendoe Pebbly Schist'. Parson noted the presence of similar lithologies within a fault-bounded wedge south of Fort Augustus. This outcrop may be traced south-westwards from Culachy [376 064] to the southern boundary of the district. Here it is contiguous with the Glen Buck Formation of the type area within the Glen Roy district (Sheet 63W). The Glen Doe Pebbly Schist and Culachy Schist are broadly contemporaneous facies variants (Parson, 1982). Here we assign both to the Glen Buck Pebbly Psammite Formation.

Within the type area, rocks of the Glen Buck Pebbly Psammite Formation structurally underlie the Grampian Group (Phillips et al., 1993; Key et al., 1997). A zone of ductile shearing marks the boundary between the two successions, the Eilrig Shear Zone (Phillips et al., 1993). This structure, traced to the southern margin of the district, outcrops within the Invermoriston district. Parson (1982) noted the sharp linear form to the eastern boundary of the Culachy outcrop, but did not recognise either the presence of mylonites or any increase in strain. The Glen Buck Formation, which here attains a maximum thickness of 600 m, is progressively cut out to the north-east by the Glen Buck Fault.

The formation appears to continue to the north-east of Borlum [387 084], initially within a narrow outcrop subparallel to the line of the Great Glen Fault. At Loch Tarff [425 100] the outcrop broadens because of late folding. The thickness of the formation here is more than 2 km. Here also the contact between the formation and the Grampian Group is a zone of ductile shearing. To the east, the outcrop is cut out by the Stratherrick–Loch Mhor Fault. Sinistral movement on the fault has displaced the outcrop by 6 km to the north-east. To the north-east, the outcrop is truncated against the south-

western margin of the Foyers Plutonic Complex. However, within the adjacent Foyers district (Sheet 73E) the formation forms roof pendants and an enclave 'ghost stratigraphy' in the Foyers Plutonic Complex. Similar pebbly psammitic rocks outcrop to the north-east of the pluton, within a tectonically bounded antiformal window (Highton, 1986). These weakly deformed non-gneissose psammites are there juxtaposed against gneissose rocks of the Central Highland Migmatite Complex.

The formation consists, predominantly, of medium- to coarse-grained arkosic psammitic rocks, with micaceous psammite and minor semipelite. The psammites contain horizons between 20 and 100 mm thick, but locally up to 0.5 m, in which lithic clasts are abundant. Pebbly horizons are laterally impersistent, traceable along strike for only a few metres. Their margins are conformable with the lithological layering in the psammites, which in areas of low subsequent strain is considered to be bedding. Within both outcrops, however, the principal foliation in the rocks is either a composite S_0/S_1 or S_1/S_2 tectonic fabric. Mostly the clast-bearing layers alternate with layers in which clasts are less common or absent. In areas of low strain within the Loch Tarff outcrop, bedding and little modified sedimentary structures, including cross-bedding, convolute and trough cross-bedding, are preserved, e.g. near Lochan Màm-chuil [439 114]. Thin, but discrete, bed-parallel bands of heavy mineral grains are present at the base of most quartzose layers, with small lenticles forming lag deposits in the cross-stratified horizons. Individual beds preserve normal graded bedding from coarse bases to pelitic tops, e.g. on the east side of Glen Tarff [395 087]. The formation is SE-dipping. Younging evidence suggests a right-way-up, upward-coarsening sequence. Laterally persistent conglomerate horizons are most common within the youngest beds exposed, e.g. on Borlum Hill [397 088]. The pebble beds are mainly indicative of active channel deposition. A poorly bedded pebbly wackestone north of Murligan Hill †415 113], with ellipsoidal clasts up to 150 mm long (Parson, 1982), may suggest minor turbidity sedimentation in the sequence.

The psammitic rocks are characteristically inequigranular, medium- to coarse-grained, grain- or matrix- supported, immature arkosic quartz-arenites. Gritty/pebbly beds are most common in the Loch Tarff area, but appear to decrease in abundance in the higher structural levels of the Culachy outcrop. Where subsequent strains are low, clasts vary from single crystals to small pebbles, mostly less than 10 mm in diameter. They comprise subrounded crystals or ellipsoidal aggregates of quartz and/or feldspar. Plagioclase compositions are limited, ranging from albite to sodic oligoclase. The Culachy outcrop and older part of the Loch Tarff area succession contain a heterogeneous clast population (Parson, 1982). This includes rock fragments of granitic composition (S 62796, 62799). The textures within the clasts are entirely metamorphic. With increased strain the clasts become increasingly ellipsoidal in shape. Recrystallisation is conspicuous, with subgrain development and suturing of grain boundaries. The terminations of the ellipsoids are ragged. These pass into the strain shadow areas where crystallisation of finer-grained quartz, feldspar and muscovite has occurred (S 60805, 62795, 62800). At outcrop this overemphasises the size of the clasts, and therefore must cast doubt on reported axial ratios (Parson, 1982). The diversity of clast type decreases markedly in successively younger beds of the Loch Tarff succession, where quartz pebbles are predominant (Parson, 1982).

The matrix consists essentially of granular quartz and feldspar, with muscovite, biotite, epidote, opaques, sphene, allanite, zircon, apatite and tourmaline. The grain size of the matrix varies from very fine- to fine-grained (< 0.2 mm), and may reflect original heterogeneity within the protolith (S 62796). Evidence from areas of low strain indicates extensive recrystallisation, with textures in the matrix largely granoblastic. Some phases within the accessory mineral assemblage are probably of detrital origin e.g. allanite, zircon, apatite and tourmaline (S 62796). Sphene overgrows and replaces ilmenite (S 62799), while growth of subhedral to euhedral pyrite crystals, the remaining opaque phase, postdates all fabrics in these rocks. Epidote, abundant in the youngest beds of the formation (Parson, 1982), is of metamorphic origin. Euhedral epidote porphyroblasts overprint both the S_1 (S 62799) and S_2 schistosities (S 62798). Muscovite is abundant in these rocks, but is of syn- to post-D_2 age.

Large (1–15 mm) salmon pink K-feldspars (mostly microcline, with subordinate perthite and mesoperthite) are common throughout the Glen Buck Formation. Previous interpretations have considered the feldspars to be detrital grains (Parson, 1982; Key et al., 1997). However, in areas of low strain, the euhedral to subhedral, untwinned growth rims of feldspar enclose rounded or ellipsoidal cores (S 62800 and 62793). Here also feldspars overgrow both bedding and the S_1 foliation in the rocks, commonly at a high angle. This suggests that the K-feldspar population is, at least in part, metamorphic, rather than detrital. With subsequent strain the

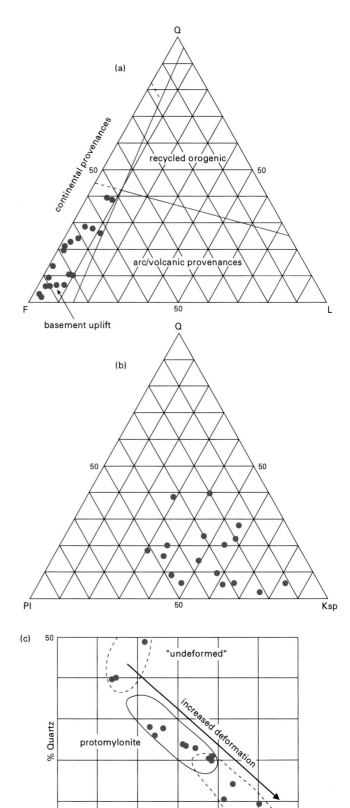

Figure 13   Mineralogy of the Glen Buck Pebbly Psammite Formation.

(a) Modified quartz (Q)–plagioclase (Pl)–lithics (L) ternary diagram (after Dickinson and Suczek, 1979). (b) Quartz (Q)–plagioclase (Pl)–K-feldspar (Ksp) diagram. (c) Plot of modal % quartz clasts versus % feldspar clasts illustrating the decrease in preserved quartz porphyroclasts with increased ductile deformation (mylonitisation). (From Phillips, 1992).

feldspars deform, and are in part wrapped by the S_2 muscovite fabric, with some evidence of recrystallisation. The abundance of K-feldspar in these psammitic rocks is consistent with their being derived from an immature protolith.

Modal data from both undeformed and mylonitic rocks (Figure 13) reflect the abundance of the two feldspars in these psammitic lithologies. The preponderance of an unstable feldspar-rich assemblage and the range of accessory minerals present suggest derivation from an uplifted metamorphic basement source area with rocks of predominantly granitic composition. The coarse clastic nature of the deposits indicates a limited period of transport, with the sedimentation probably close to the source area as braided stream type deposits. Systematic studies of sediment transport direction have not been undertaken, and the location of the source area is consequentially not known.

Correlation of the Glen Buck Pebbly Psammite Formation of the Invermoriston district with the Beinn Mheadhoin Pebbly Psammite Formation, north-east of the Foyers Plutonic Complex (Highton, 1986), is unequivocal. The Gairbeinn Pebbly Semipsammite Formation (Haselock et al., 1982) of the Glen Roy and Dalwhinnie districts (Sheets 63W and 63E) may also be a lateral equivalent. This correlation has not yet been proved. Lithological differences between the formations are minor. However, conglomeratic horizons are absent in the Gairbeinn Pebbly Semipsammite Formation. There is a striking difference in the apparent grade of metamorphism. Within the Invermoriston and Glen Roy districts, the Glen Buck Formation is at greenschist facies (Key et al., 1997). Kyanite is present in semipelitic rocks underlying the Gairbeinn Pebbly Semipsammite Formation on Gairbeinn, at the eastern edge of the Glen Roy district (Haselock et al., 1982). This suggests lower to middle amphibolite facies conditions. The occurrence of a persistent tectonic break, or breaks, above the formations prevents the establishment of a clear stratigraphical link through to the structurally overlying Grampian Group. The Glen Buck Pebbly Psammite Formation is separated from the Moine rocks of the Northern Highlands by the Great Glen Fault Zone. There are many similarities, in terms of lithological composition and sedimentary environment, with rocks of the Morar Group of the Moine Supergroup. Therefore a tentative correlation between rocks of the northern and Grampian Highlands might be possible (Harris et al., 1981).

Structure

D_1 STRUCTURES

Rocks of the Glen Buck Pebbly Psammite Formation have a similar polyphase history of deformation to that of the overlying Grampian Group (Parson, 1982). Bedding within the Loch Tarff outcrop dips essentially to the south-east, with older beds lying to the north-east adjacent to the Foyers Plutonic Complex. The earliest phase of deformation in these rocks is manifest as an irregular spaced cleavage in the psammitic lithologies or biotite (mimetically replaced by muscovite) in the semipelitic interbeds. Cleavage planes within the psammites, spaced at intervals of approximately 10 mm, are either laterally impersistent or anastomosing. To the north-east of Lochan Màm-chuil [436 113], the S_1 fabric dips at approximately 50° to the south-east, cutting bedding at a moderate angle (Parson, 1982). The fabric is folded around later D_2 folds. Early folds have not been recognised within rocks of the formation. Clasts within the gritty/pebbly layers become oblate in form, lying within this fabric.

Bedding and the S_1 fabric relationships are poorly defined within the Culachy outcrop. They are subparallel. Clasts are predominantly oblate in form, flattened within the fabric.

D_2 FOLDS AND SHEAR ZONES

An upward-facing, intermediate-scale synform/antiform pair of D_2 age controls the distribution of lithologies within the Loch Tarff outcrop. These upright, NNE-trending structures plunge at 30–50° to the south-west (Parson, 1982). The antiformal structure is tight, with locally abundant minor parasitic folds. The synform is tight to isoclinal in form. A spaced S_2 axial planar crenulation cleavage is present in the fold cores. On the limbs of the folds, this transposes the earlier fabrics into an intense composite mica fabric. To the west of Murligan Hill [4070 1002], semipelitic interbeds within the arkosic sequence contain small (< 0.3 mm) euhedral garnet porphyroblasts (S 62804). The garnets are mostly inclusion-free, single-stage porphyroblasts. Growth appears to be syn-tectonic. The larger porphyroblasts contain S-shaped inclusion trails defined by fine-grained opaque minerals. The interbeds have some similarity to rocks at the base of the Grampian Group in Glen Buck (Phillips, 1992), and may represent tectonic intercalations. Clasts in the psammitic lithologies are wrapped by the S_2 fabric. Crystallisation of quartz, feldspar and orientated flakes of muscovite occurs in the pressure shadow areas, resulting in a bearded texture to the overgrowths.

A tight synformal fold is present within in the Culachy outcrop (Parson, 1982). The structure closes to the NE, but plunges shallowly to the SSE. A spaced S_2 cleavage fans around the closure. There are few reliable way-up structures here, although graded-bedding north-west of Liath Dhoire [364 038] suggests that the sequence is right-way-up, and upward-facing.

In the Loch Tarff outcrop there is a significant increase in the strain gradient towards the boundary with the overlying Grampian Group. The contact, where exposed on Borlum Hill [397 088] and Murligan Hill [412 103], is a zone of mylonites and phyllonitic proto-mylonites (Phillips, 1992). Within the zone, previously oblate pebbles have recrystallised, with increasing deformation, to prolate forms. Axial ratios of 70:5:1 are not uncommon adjacent to the zones of mylonitic rocks. The high-strain zone is characterised by quartz ribbon fabrics, and the development of tabular subgrains (S 62801, 62803). A linear extension fabric, defined by the deformed clasts and elongate quartz, has a consistent shallow to moderate plunge to the south-east. Although the contact between the Glen Buck Formation and

Grampian Group is not exposed to the south of Culachy, a linear extension fabric is present throughout the outcrop. Kinematic indicators from within the high-strain zone, e.g. S-C fabrics and deformed porphyroclasts have a consistent north-west-directed sense of shearing. This suggests the emplacement of the Grampian Group over the underlying rocks of the Glen Buck Formation.

Metamorphism

Semipelitic lithologies are a minor component of the Glen Buck Formation, and index minerals are generally absent in these rocks. The mineral assemblage: muscovite + chlorite + plagioclase ± K-feldspar is predominant in rocks from the Culachy outcrop and the area of Borlum Hill. In the Glen Roy district, this assemblage is thought to represent the peak of prograde, greenschist facies metamorphism (Key et al., 1997). There is an apparent jump in metamorphic grade across the Eilrig Shear Zone to amphibolite facies assemblages of the Grampian Group. This is interpreted as evidence for a major displacement on this structure (Phillips et al., 1993; Key et al., 1997).

The presence of garnet porphyroblasts within semipelitic interbeds of the Loch Tarff succession (S 62804) shows higher-grade metamorphic conditions here. However, their tectonostratigraphical position is, at present, uncertain. The peak of metamorphism, at upper greenschist to lower amphibolite facies, was probably syn-D_2, but predated the shearing event. A later overprint or post-shear re-equilibration is shown by the mimetic replacement of garnet by biotite. Anhedral grains and aggregates of epidote are intimately associated with these sites. In the phyllonitic rocks of the shear zone, euhedral epidote porphyroblasts overgrow the shear fabrics (S 60804, 62800). Chlorite is a secondary, retrogressive mineral, replacing both biotite and garnet. It is abundant in rocks within and adjacent to the Great Glen and Glen Buck faults, where veining by cataclasite and carbonate is common. Phillips (1992) obtained temperatures of 250–350°C on these rocks, based on two-feldspar thermometry and Al in chlorite.

ROCKS WITHIN THE GREAT GLEN FAULT ZONE

Rocks within the Great Glen Fault Zone occur in a narrow fault-bounded outcrop that extends from near Borlum [385 082] south-westwards to the southern margin of the district. They are continuous with similar fault-bounded rocks that have been traced through the Glen Roy district as far as Loch Lochy (Key et al., 1997).

The rocks are mainly psammitic, intensely fractured and crushed as result of movements within the fault zone. The original foliation of the metasedimentary rocks is not preserved in many exposures, and the main fabric recognised in the field is that of numerous dislocations and slip planes on all scales and orientations.

The rocks are generally not well exposed but form a series of hills south-west of Fort Augustus, of which Meall a' Cholumain [361 049] reaches a height of 315 m above OD. The rocks are well exposed in Glen Tarff and in the valley of the Calder Burn [348 028], outwith the district to the south.

Lithology

The main rock type is a fine- to medium-grained psammite, consisting of quartz, plagioclase, K-feldspar and a subordinate amount of white mica, chlorite, epidote and very fine-grained opaque oxides. The mafic minerals occur in thin layers, up to 2 mm thick, more easily recognised in thin section than in the field, which define a foliation, possibly a metamorphosed bedding structure.

Semipelitic rocks are uncommon, outcropping near the southern margin of the district only. The rocks are gneissose, with quartzofeldspathic segregations up to 75 mm thick. Thin intercalations of psammite are common. Chlorite, and less commonly muscovite, replace biotite in the melanosomes. The gneissose fabric, defined by parallel flakes of mica, is concordant with the lithological layering. Parson (1982) recorded the presence of dark green schistose amphibolites in these metasedimentary rocks, occurring as thin discontinuous concordant sheets and lenses. Granitic and pegmatite veins occur throughout the outcrop area and exhibit both concordant and cross-cutting relationships with the foliation in the metasedimentary rocks.

Lithostratigraphical correlation

The metasedimentary rocks within the Great Glen Fault Zone are separated from other mapped lithological units by faults of unknown displacement. Net displacement along the Great Glen Fault Zone are equivocal (but see Johnstone and Mykura, 1989) and direct stratigraphical correlation across it is not possible. The rocks have no distinctive lithological features, the chief rock type being a mica-poor psammite similar to lithological units recognised on both sides of the Great Glen. The presence of foliated amphibolite bodies, a gneissose fabric and granitic veins suggest a correlation with either rocks of the Moine to north-west of the Great Glen or the Central Highland Migmatite Complex.

FIVE

Pre-Caledonian igneous rocks

The tectonostratigraphical succession of the Northern Highland Moine contains many suites of meta-igneous rocks. These represent the emplacement of both acid and basic magmas prior to the peak of the Precambrian tectonothermal event in the Moine. A suite of gneissose granite bodies, collectively referred to as the West Highland Granite Gneiss, represent the earliest intrusive event within the Sgurr Beag Nappe (Barr et al., 1985). These crop out mainly close to the Glenfinnan–Loch Eil group boundary between Ardgour and Loch Cluanie, although to the east of Loch Quoich their emplacement appears to have been at higher levels in the stratigraphical sequence. Suites of metabasic rocks, with mostly tholeiitic composition, mark the extensive emplacement of differentiated sill complexes (Winchester, 1976; Rock et al., 1985) which postdate the granite magmatism. Both the gneissose granite and metabasic rocks record all the tectonic events recognised in their metasedimentary hosts. However, the age of their emplacement is equivocal. Barr et al. (1985) considered the gneissose granite protoliths to be a series of S-type granitic sheets (cf. Chappell and White, 1975). Emplacement was into country rocks already undergoing middle to upper amphibolite facies metamorphism, contemporaneous with the earliest recorded tectonic event (D_1). Brook et al. (1976) obtained an Rb/Sr isochron from the most southerly Ardgour Granitic Gneiss mass. This gave an apparent age of 1028 ± 43 Ma with an initial Sr^{87}/Sr^{86} ratio of 0.709. Other methods have not verified this age (Aftalion and van Breemen, 1980), and more recent work indicates that granite emplacement and pegmatite segregation occured at c. 880–900 Ma (Rogers et al., 1995). Thus the Rb/Sr study must, therefore, remain questionable (Sanders et al., 1984; Strachan and Treloar, 1985).

Pre-Caledonian metabasic rocks are also present within the Precambrian sequence to the south of the Great Glen Fault, but none are recorded within the Invermoriston district.

WEST HIGHLAND GRANITIC GNEISS

Within the Invermoriston district a major, apparently sheet-like body of gneissose granite crops out to the south-west of Fort Augustus. Other minor bodies occur in Glen Moriston west of Dalchreichart, e.g. [285 135]. All lie entirely within the outcrop of the Loch Eil Group. The age of these bodies is unknown, although they are probably contemporaneous with the Ardgour mass.

FORT AUGUSTUS

The gneissose granite at Fort Augustus (Figure 14) forms an elongate NE-trending outcrop of some 35 km², bounded to the south-east by the Great Glen Fault.

Exposure is variable, with much of the outcrop concealed by superficial deposits. The gneiss is typically equigranular, medium grained and pinkish white weathering, with local developments of coarse-grained pegmatitic or granitic segregations. Only upper surface contacts of the mass are seen. These are sharp, and concordant with the regional foliation in the Moine. On the regional scale, however, the mass transgresses the local stratigraphy, but at a low angle. Within areas of comparatively low D_2 strain, e.g. to the north-west of Meall Mór [313 043], interdigitation of the gneiss and host rocks may reflect an original, though tectonically modified, emplacement contact. The general form of the gneissose granite body defines a broad antiform of (?)D_3 age, controlled by impersistent gentle to open folds, whose style and amplitude are similar to those affecting the host metasedimentary sequence.

The predominant foliation throughout much of the outcrop is a composite S_1/S_2 fabric defined by orientated micas, biotite-rich laminae, elongate mafic aggregates and quartzofeldspathic segregations. The foliation is mostly shallowly dipping to the west or north-west. Locally, it becomes steeply inclined in the hinges and short limbs of reclined D_2 folds, e.g. near Coille Réidh nan Làir [322 053] and in Auchteraw Wood [339 069]. Where D_2 strain is low, e.g. near the latter locality, the rocks have the appearance of homogeneous granite. Veins of coarser-grained leucogranite and pegmatite crosscut this fabric. Thin selvedges of biotite and accessory minerals enclose these segregations. These are coeval with the regional (D_1) gneiss-forming event in the host rocks (Barr, 1983; 1985; Barr et al., 1985). The S_1 foliation is generally moderate to steeply inclined. This fabric, with the quartzofeldspathic veins, defines tight, small- to intermediate-scale D_2 fold closures. These are prominent on the short limb of the larger D_2 structures, with a discrete S_2 axial planar mica fabric. Throughout much of the outcrop, the early coarser-grained granite and pegmatitic segregations have become transposed into the S_2 fabric. Locally, e.g. at Torr a' Choiltreich [3693 0757], small subhorizontal shear zones modify the S_1/S_2 fabric. The shears mostly propagate on the limbs of the minor folds. Displacements are relatively minor, about 0.2 m, and record a general north-westerly sense of overthrusting. Localised partial melting accompanied the shearing event, e.g. near the last locality [3700 0769]. Small segregations of granitic material occur in the shears (S 72180, 72197, 72234), but also extend away from the zones as small granitic or pegmatitic veinlets.

The gneiss is essentially monzogranitic, comprising varying proportions of quartz, plagioclase and K-feldspar (perthitic microcline), with biotite (up to 7 per cent) and minor muscovite. Textures are allotriomorphic to granoblastic, with grain sizes mostly less than 2 mm, but up to

Figure 14 West Highland Granitic
Gneiss near Fort Augustus.

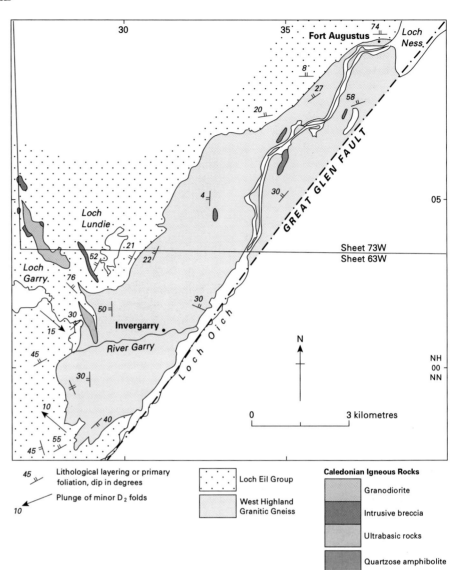

4 mm in the pegmatitic segregations. Plagioclase (An_{25-17}) is
mainly metamorphic. Some crystals
with subhedral form preserve
patchy zoning (S 72230) and may
be of relict igneous origin. Plagioclase laths, enclosed poikilitically by
microcline, show extensive resorption. Myrmekite is common. Garnet
is ubiquitous, forming small
(< 0.5 mm), elongate, poikiloblastic
crystals with inclusions of quartz,
sphene and allanite (S 72199,
72230). Garnets show evidence of
disequilibrium, and are partly
replaced by quartz-plagioclase-bio-
tite symplectites (S 72230). Accessory minerals, including apatite,
zircon, sphene and opaque minerals with minor allanite and monazite, are found mainly within the
mica-rich laminae and selvedges to
the coarser-grained segregations.
Zircon, seen mostly as inclusions in
biotite, shows a variety of morphologies, with indications of possible
restite or inherited components.
Most of the gneissose granite is
fresh. However, minor zones of
alteration (c. 200 m wide) are present adjacent to the Great Glen
Fault, where the rocks become reddened (S 72199).

Small diffuse patches of tonalitic
composition occur within the gneiss
near the Invervigar Burn [3300 0600, 3362 0370, 3360
0402]. They contain a dark green amphibole, large (up to
4 mm) euhedral crystals of sphene and feldspar resorption
textures (S 72198). The D_2 fabric is recognisable, picked
out by small aggregates of biotite and amphibole. However,
this is mimetic and larger amphibole crystals overprint the
fabric. The patches are not considered to reflect hybridisation within the granite protolith. However, they are contemporaneous with a pre- to syn-D_3 alkali igneous event,
which is locally manifest in the emplacement of quartzose
amphibolite plugs (see Chapter 6).

GLEN MORISTON

Gneissose granite forms an outcrop traceable northwards
for 5 km from Balintombuie [282 130] and southwards
to Achlain [279 123], where it is exposed in a cutting
resulting from the realignment of the A 887 (Figure 3).
The gneiss forms a concordant sheet, about 40 to 50 m
thick, dipping eastwards at an angle of 20°. A thinner,
discontinuous sheet occurs on the hillside to the south of
Achlain [280 108]. The lower contact of the main sheet

outcrops within the Allt Baile nan Carn [2757 1640] and
its tributary [2740 1718]. Some 150 m to the east, the
stream turns sharply to the south to follow the upper
contact. This is a fault plane dippings to the east at 40°,
approximately concordant with the foliation in the gneiss.
The rocks close to the fault are shattered and reddened.

The Glen Moriston gneissose granite is similar in
appearance to the major occurrence at Fort Augustus. It
is greyish to pinkish in colour, with a well-developed
foliation and contains many quartzofeldspathic segregations, each with a narrow biotite-rich selvedge (Plate 3).
The foliation of the gneiss is mainly an S-fabric. It forms
from the parallel orientation of biotite, but there are also
micaceous laminae that allow the rock to split into slabs.
Lenticular quartzofeldspathic segregations also con-
tribute to the foliated structure. The mica fabric is a D_1
structure, folded during D_2 (Figure 7). A new axial planar
schistosity is present in the hinge zones of the folds. On the
fold limbs, the fabric is a composite D_1–D_2 structure.

An unusual hornblendic variety occurs in a zone about
3 m thick exposed in the Allt Bail' an Tuim Bhuidhe

Plate 3 Concordant low-angled, easterly dipping sheet of gneissose granite in rocks of the Upper Garry Psammite Formation. Leucosomes with enclosing biotite-rich selvedges prominent to top of exposure. Cut by thin sheets of schistose amphibolite. Road cutting on the A887 near Achlain [278 124]. (D 5070)

[2851 1387]. Hornblende is disseminated through layers a few centimetres thick and also occurs in the quartzofeldspathic segregations where stumpy crystals up to 10 mm long are prominent.

Geochemistry

The small dataset (Table 2) from the Fort Augustus body suggests a restricted compositional range across the outcrop, with minor variations only in major or trace element contents. All samples from the main part of the mass are peraluminous, alkali-calcic granites (*ss*), with high Fe (total Fe as FeO) relative to Mg (Fe/Fe+Mg = 0.8–0.87). The rocks are all corundum normative, with a restricted range of composition (69–72 wt.% SiO_2) and an average A/CNK ratio of 1.14. Components plot slightly away from the cotectic line in the Quartz-Albite-Orthoclase system (Figure 15), towards the Quartz + Plagioclase + Liquid + Vapour cotectic surface (Barr et al., 1985).

Trace elements show a modest enrichment of HFS elements in the granites (Figure 16). The elevated Zr content (600–700 ppm) in these rocks is unusual, given the limited solubility of Zr in peraluminous melts (Watson and Harrison, 1983). This suggests that the behaviour of Zr in the gneissose granite is in part controlled by the presence of restite zircon. In keeping with other components of the West Highland Granitic Gneiss (Barr et al., 1985), the Fort Augustus rocks have apparent S-type characteristics (Y+Nb> 80, Rb> 100).

Table 2 Whole-rock analyses of samples from the West Highland Granitic Gneiss near Fort Augustus.

Sample no. Grid ref.	S 72189 3650 0754	S 72230 3343 0453	S 72234 3466 0665	S 72199 3283 0602
SiO_2 (wt.%)	71.07	71.08	69.04	72.05
TiO_2	0.69	0.65	0.82	0.64
Al_2O_3	12.88	12.68	13.34	12.90
FeO*	4.91	4.60	5.48	4.02
MnO	0.06	0.05	0.07	0.11
MgO	0.77	0.69	0.93	0.64
CaO	1.91	1.79	2.28	1.81
Na_2O	3.15	3.03	3.70	3.07
K_2O	4.00	4.12	3.25	4.42
P_2O_5	0.15	0.15	0.18	0.09
Loss on ignition	0.43	0.59	0.42	0.71
Total	100.02	99.53	99.51	100.46
Zn (ppm)	68	59	61	69
Rb	166	198	189	199
Sr	166	148	156	153
Zr	635	636	701	598
Y	73	88	82	88
Nb	15	13	16	15
La	70	110	110	90
Ce	20	50	60	40
Ba	750	620	490	630
Pb	20	18	19	23
Th	15	19	21	16
U	3	6	5	6
CIPW norms				
Qz	29.10	30.27	25.52	30.16
Ab	26.73	25.88	31.56	26.08
An	8.52	7.98	10.22	8.29
Or	23.71	24.58	19.36	26.22
Cndm	0.27	0.35	0.03	0.05

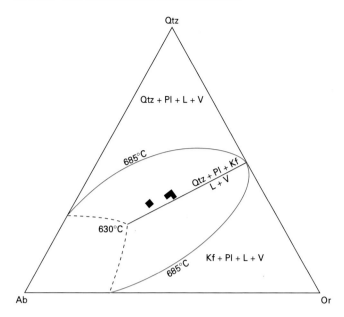

Figure 15 Plot of normative quartzofeldspathic components in the West Highland Granitic Gneiss near Fort Augustus, projected onto the quartz (Qtz)– albite (Ab)–orthoclase (Or) face of the salic tetrahedron. Projections of the Qtz + Pl + Kf + L + V and Qtz + Pl + L + V cotectic surfaces are plotted for P_{H_2O} = 7 kb (after Winkler, 1979).

METABASIC ROCKS

The metabasic rocks are amphibolite and schistose amphibolites. They are common within the rocks of the Northern Highland terrane and fault-bounded gneissose lithologies of the Great Glen Fault Zone (Figure 17). Bodies vary in size from layers a few centimetres thick, to pods and sheets many metres thick. Their distribution within the Moine outcrop is uneven. However, they are particularly abundant in a belt running north-east from Fort Augustus, and in the Cannich–Glen Affric area. Amphibolite and schistose amphibolite are not present within either the Glen Buck Pebbly Psammite Formation or Grampian Group rocks of the district.

Many amphibolite bodies crop out on the hillside above Achnaconeran, to the north of Invermoriston (Figure 18). The larger bodies are more resistant to erosion and give rise to prominent ice-moulded crags. Contacts with Moine host rocks are are rarely seen. Within psammitic lithologies, bodies are mostly parallel-sided sheets up to 20 m thick. They are concordant with the regional foliation in the host rocks. The larger bodies extend for several hundred metres before either tapering out or ending abruptly. In the semipelitic or mixed lithology units, many bodies are podiform, e.g. in the Allt Saigh [4268 1917]. These may commonly reach 15 m in thickness, with length to width ratios of three to one or less. The larger pods are foliated only at their margins, while the smaller bodies (< 1 m) are schistose throughout. The latter commonly occur as trails of lenticles following the country rock foliation that adapts to them, e.g. at [4260 1817]. The

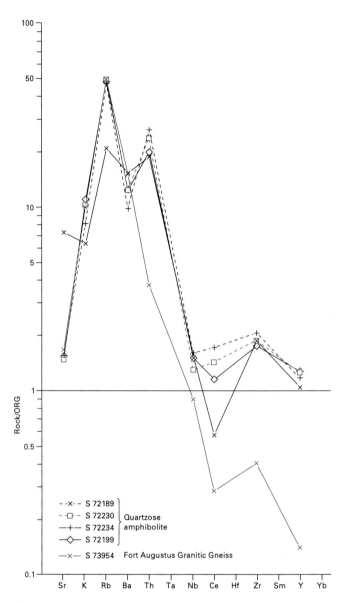

Figure 16 Multi-element variation diagram of selected samples of the West Highland Granitic Gneiss near Fort Augustus, normalised to Ocean Ridge Granite (after Pearce et al., 1984).

trails are the result of boudinage of formerly continuous sheets, with extension mostly parallel to the S_0/S_1 foliation (see Plate 1). Disruption of the protolith intrusions occurred mainly during the D_1 event. There is, however, evidence of further boudinage in D_2. Most of the sheets appear to have behaved as rigid masses during D_2 folding. In places they are quite discordant, e.g. the body east of Achnaconeran [4202 1830] crosscuts a D_2 fold on a tectonic contact (Figure 19a). At Invermoriston Bridge [4197 1655] metabasic sheets are not only separated into pods but also folded (Figure 19b). The pods are generally massive to weakly foliated amphibolite in their cores, which pass into schistose rims. Here, biotite almost completely replaces the amphibole. The schistose margins, with branches into the psammitic country rock, define D_2

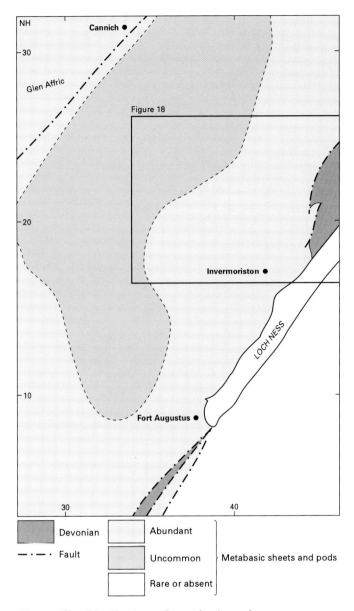

Figure 17 Distribution of metabasic rocks.

shear zones dividing the rock mass into a complex of displaced lenticles. The metabasic bodies exposed in Allt Baile nan Carn [27 13] are mostly concordant sheets of schistose amphibolite. They commonly reach 15 m in thickness, and are garnetiferous throughout. The penetrative schistosity in these bodies is parallel to the sheet margins. Where folded during D2, this presumably D1 fabric is crenulated. Within the Fort Augustus gneissose granite, trains of amphibolite boudins occur at several localities, e.g. Torr a' Choiltreich [3700 0769], [3720 0819] and Torr Dhùin [3470 0685]. These bodies show angular discordances with the gneissose foliation and their margins are intensely schistose.

Rock et al. (1985) describe the field relationship of the metabasic bodies in the Cannich to Glen Affric area. They achieve a maximum outcrop density at Eas Maol Mhairi in the River Cannich [3172 3237, 3197 3229]. Variations in the size and fabric of the bodies are a function of country

rock competence during deformation. The smaller, disrupted amphibolites and schistose amphibolites occur within pelitic lithologies.

A D_1 schistosity is present in most metabasic bodies. The igneous phase of their evolution predates the D_1 deformation. It is likely that the sheets and pods are the dismembered remnants of an original sill complex of great lateral and vertical extent.

Petrography

The metabasic bodies range from coarse-grained, massive rocks (amphibolite), with almost no trace of igneous texture, to schistose rocks (hornblende-schist and biotite-schist). The large amphibolite bodies show only slight macroscopic variation; all are coarse grained and garnet is only present in the schistose margins. Primary igneous layering and internal contacts suggesting multiple intrusion are seen at only one locality, beneath the bridge at Balnacarn [2731 1312]. Here a sheet at least 4 m thick comprises a coarse-grained facies separated from a fine-grained facies by a sharp contact. Xenoliths, even of the local country rock, are uncommon.

In thin section the amphibole is pale brownish green in colour; analyses show it to be edenitic to pargasitic in composition. The crystals have ragged outlines and contain rounded inclusions, mainly of quartz. The amphibole in the schistose amphibolites is mid to pale green in colour and dominantly magnesio-hornblende. In the biotite-schist occurring along the tectonised margins of the amphibolite bodies, biotite almost completely replaces hornblende. Aggregates of green hornblende, with minor biotite in a few samples, rim clinopyroxene (salite).

Plagioclase (calcic oligoclase to andesine) forms granoblastic grains, very commonly with narrow but distinct sodic rims. Most of the samples contain clinozoisite which is intergrown with or closely associated with plagioclase. The clinozoisite content is inversely proportional to plagioclase, indicating that it has replaced the feldspar. Quartz and biotite, commonly chloritised, are very variable in abundance. Garnet is scarce.

Almost all the thin sections show opaque minerals, probably ilmenite, rimmed by sphene. These often form elongate aggregates and trails, aligned parallel to the schistosity. Sphene generally forms inclusion trails in the amphibole, which suggests that recrystallisation and replacement of ilmenite took place before the final crystallisation of hornblende.

Geochemistry

Table 3 lists representative analyses of metabasic rocks from the district and additional data by Rock et al. (1985). Variations in the data suggest some mobilisation of LIL elements. This occurred either on emplacement or during their subsequent metamorphism. The coherence in correlations of HFS elements (mostly incompatible in basaltic systems), e.g. TiO_2 and Zr (Figure 22), probably represents primary magmatic processes (Pearce et al., 1974; Winchester and Floyd, 1976).

The absence of any primary mineralogies in these rocks constrains any satisfactory evaluation of the proto-

Figure 18
Distribution of metabasic rocks north of Glen Moriston. Exposure is generally sufficient to imply that the mapped distribution approximates the real distribution.

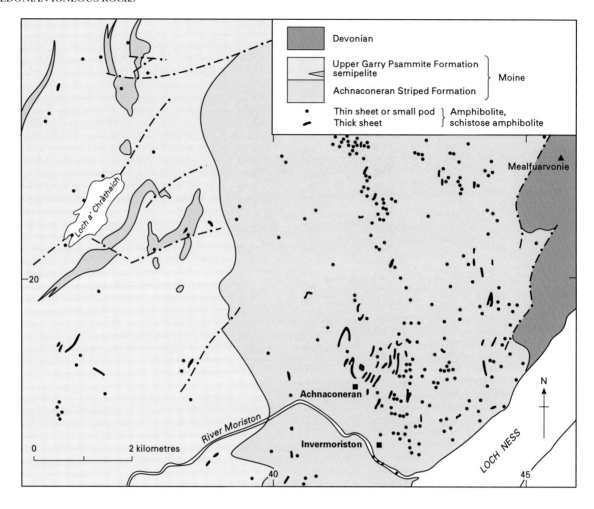

Figure 19 Structural relationships of amphibolite masses. (a) Vertical section showing amphibolite cutting a D_2 fold in micaceous psammite. A biotite-rich schistose selvedge, 0.1 m thick, lies along the tectonic contact near Achnaconeran [4202 1830].
(b) Vertical section showing folded (D_2) metabasite sheet. A core of massive to weakly foliated amphibolite is enclosed by biotite-schist with branches extending into the country rock. The biotite-schist defines a system of D_2 shear zones. Invermoriston Bridge [4197 1655].

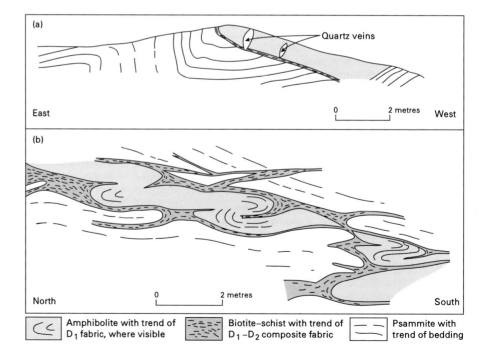

Table 3 Representative whole-rock analyses of metabasic rocks.

Sample no. Grid ref.	S 67246 3824 1821	S 67243 4260 1817	S 72404 4187 1788	S 67241 4145 1830	S 72391 3401 1868	S 72393 3645 1813	S 71340 2915 0520	S 71329 29240 502
SiO_2	47.60	48.30	47.90	48.70	48.50	48.20	46.80	50.51
TiO_2	1.47	1.72	2.60	1.86	2.75	2.50	3.78	3.20
Al_2O_3	16.86	16.61	14.00	15.14	14.31	14.36	12.60	12.11
FeO^*	10.77	11.26	14.65	12.68	13.71	13.98	16.84	16.97
MnO	0.20	0.20	0.21	0.20	0.20	0.21	0.31	0.25
MgO	9.10	8.20	8.04	7.59	7.06	6.79	6.36	5.92
CaO	11.53	11.56	10.62	10.41	11.60	11.54	8.59	9.67
Na_2O	1.23	1.38	1.66	1.39	1.75	1.93	2.50	0.81
K_2O	1.82	1.48	0.83	1.81	0.80	1.07	1.64	1.16
P_2O_5	0.16	0.16	0.29	0.15	0.23	0.23	0.31	0.42
Total	100.74	100.87	100.80	99.93	100.91	100.81	99.73	101.02
V	nd	nd	nd	nd	nd	nd	430	410
Cr	328	170	204	172	151	108	100	70
Co	47	49	50	50	50	56	40	36
Ni	129	84	66	54	50	48	50	36
Cu	7	6	9	8	79	209	21	56
Zn	80	82	113	90	93	98	125	133
Rb	89	58	32	63	27	28	58	60
Sr	163	173	153	167	181	141	208	100
Y	29	33	53	37	45	45	66	62
Zr	95	112	174	108	140	140	336	318
Nb	4	4	4	2	5	2	5	2
La	10	20	27	12	19	7	50	50
Ce	nd	nd	nd	nd	nd	nd	20	20
Ba	83	90	96	61	63	48	220	240
Pb	10	9	8	10	10	9	5	5
CIPW norms								
Qz	0.00	0.00	0.00	0.00	0.00	0.00	0.00	8.24
Ab	10.30	11.67	13.88	11.73	14.62	16.14	21.09	6.76
Or	10.64	8.74	4.85	10.67	4.67	6.25	9.66	6.76
An	34.79	34.79	28.01	29.69	28.51	27.08	18.29	25.65
Di	16.93	17.50	18.29	17.15	22.13	23.18	17.43	15.84
Hy	6.13	13.86	21.19	18.35	19.46	11.69	8.98	25.71
Ol	15.45	7.29	4.69	5.44	1.60	7.04	12.13	0.00
Mt	2.58	2.45	3.50	3.06	3.28	3.34	4.06	4.05
Il	2.77	3.27	4.89	3.53	5.17	4.70	7.17	6.00
Ap	0.37	0.37	0.66	0.35	0.53	0.53	1.15	0.96

lith compositions, and modelling the evolution of the suite. There is a modest Fe (Figure 20) and Ti-enrichment. This accompanies an increase in the concentration of HFS elements and depletion in Ni and Cr, characteristic of the evolution of tholeiitic magmas. Rock et al. (1985) report a single sample of apparent andesitic composition. The metabasic rocks are predominantly olivine normative, with few containing either quartz or nepheline in the norm. Comparison of variations in the suite with known tholeiitic suites (e.g. Macdonald et al., 1981) suggests evolution through crystal fractionation and accumulation. This would involve mainly a combination of olivine, clinopyroxene and plagioclase. The major and trace element data for the immobile elements form a continuous trend (Figure 21), consistent with a genetic relationship between the rock types. The positive correlation of MgO with both Ni and Cr argues for a probable evolution by the removal or addition of olivine and clinopyroxene. The scattered, though gradual decrease in CaO and Sr content and distinct positive correlation of Al_2O_3 argue for the involvement of plagioclase during fractionation. Other elements incompatible in olivine, e.g. P_2O_5 and TiO_2, concentrate into the more evolved rocks (Figure 21). Whether the behaviour of these elements is a reflection of apatite, titanomagnetite and ilmenite phenocrysts in the protolith magma, is uncertain.

Rock et al. (1985) indicate that the pattern of Ti, Zr, Y and P enrichment suggests that many smaller metabasic bodies represent fractionated magmas tapped from large ferrogabbroic sills. Most compositions represented have some similarity to mid-stage differentiates. For example, within the 15 m thick body exposed in the Allt Lundie [2930 0520], element variations (Figure 22) show an apparent fractionation trend; the highest exposures of the body represents more evolved compositions. Neither highly evolved differentiates nor cumulus rocks are present in the suite (Figure 22). However, the high Cr content of sample S 67246 may reflect the original presence of some cumulus minerals.

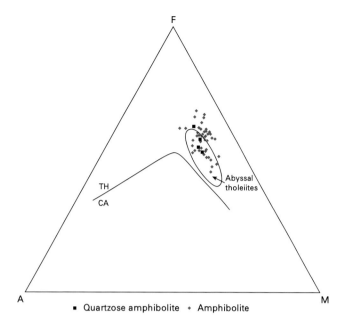

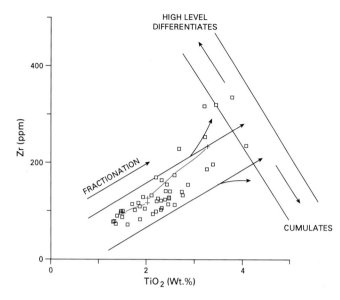

Figure 20 AFM diagram of the metabasic rocks analysed from the Invermoriston district. The line TH-CA represents the tholeiite–calc-alkaline divide (Irvine and Barager, 971). The field of abyssal tholeiites follows Miyashiro et al. (1970).

Figure 22 Variation within the metabasic rocks compared with a model fractionation scheme of Dalradian metabasic rocks (after Graham, 1976). Crosses indicate samples from within a single body north of Loch Lundie [2930 0520].

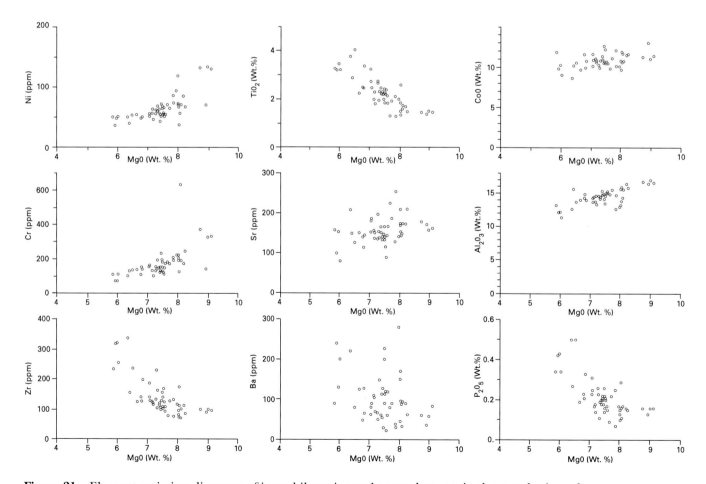

Figure 21 Element variation diagrams of immobile major and trace elements in the metabasic rocks.

SIX

Caledonian and post-Caledonian igneous rocks

The main period of intrusive igneous activity within the metamorphic Caledonides coincided with the waning stages of the Ordovician–Silurian orogenic event (c. 435 to 400 Ma). This saw the emplacement of large volumes of calc-alkaline magma into the middle and upper crustal levels. The emplacement of alkali igneous complexes, e.g. the Glen Dessary Syenite at 456 ± 6 Ma, and widespread alkali metasomatism, are contemporaneous with the early stages of the orogenic event. Within the Invermoriston district, a suite of quartzose amphibolite intrusions represents this igneous event. Granite-pegmatite vein complexes throughout the Highland area are contemporaneous with the peak of the Ordovician metamorphic event (syn- to post-D_3) in the Northern Highlands. The Glenmoriston Vein Complex represents granitic intrusions of this age within the Invermoriston district. The main period of magmatism is associated with the waning of orogenic activity. It spans the deformation associated with NW-directed thrusting, and subsequent uplift. There is a broad similarity between the plutonic and minor intrusive suites on either side of the Great Glen Fault Zone. This suggests a continuity in the magmatic history of this part of the orogen, although direct comparison is not possible. The close of the Caledonian orogeny marks a hiatus in magmatic activity, which extended until the end of the Palaeozoic. Minor magmatic activity followed during Permo-Carboniferous times, with the emplacement of a suite of camptonitic, alkali lamprophyre dykes. Their emplacement was probably consequential upon crustal extension related to the opening of the North Sea graben. Within the Invermoriston district these intrusions occur only to the north-west side of the Great Glen.

PRE-D_3 INTRUSIONS

Quartzose amphibolite

Several bodies of quartzose amphibolite occur within the outcrop of the Fort Augustus Granitic Gneiss. These were formerly interpreted as hybrid granitic migmatite resulting from partial assimilation of Moine xenoliths incorporated into the granitic protolith during intrusion (Parson, 1982). These bodies have been found to contain a heterogeneous xenolith assemblage, including fragments of gneissose granite. A foliated, medium-grained quartzose amphibolitic host encloses the xenoliths (Highton, 1994). The mineralogy, textural make-up and composition are similar to the 'hornblende-biotite-schists' reported by Rock (1984) from the White Bridge area within the adjacent Glen Roy district (Sheet 63W).

Two elliptical quartzose amphibolite masses, of approximately 1 km in length, occur within the outcrop of the gneissose granite (Highton, 1994, fig. 2). These crop out near Coiltry and on Torr Dhùinn. Several smaller bodies occur in the area of Torr a' Choiltreich. All are discordant to the regional foliation in the granitic orthogneiss host, and preserve an intense NE-trending L-S tectonite fabric.

Coiltry

Small outcrops of quartzose amphibolite occur immediately to the south of the Caledonian Canal, e.g. at Coiltry [349 060]. Contacts with the enclosing gneissose granite are not seen. Xenoliths are predominantly of metasedimentary rocks, comprising lithologies similar to the local Loch Eil Group with attendant concordant sheets of pre-D_1 amphibolite, and gneissose granite. Contacts between the xenoliths and host lithology are everywhere sharp, with no evidence of assimilation. The density of xenoliths varies from tightly packed, with little or no matrix present, to isolated fragments. The fabrics and structures preserved in the xenoliths resemble those of the regional D_2 fabrics recognised in both the Moine and gneissose granite; however, their orientations are random. All are discordant to the steeply dipping (?D_3) foliation in the quartzose amphibolite matrix; however, the long axes to the 'enclaves' lie subparallel to this fabric.

Torr Dhùinn

A similar body occurs below the remains of a vitrified fort on Torr Dhùinn [348 069] (Highton, 1994, fig. 2). Contacts are ill defined, with the host orthogneiss passing into a marginal breccia zone up to 5 m wide. This consists of a clast-supported breccia comprising angular blocks of gneissose granite, with rare metasedimentary fragments. An interclast matrix is not detectable, with boundaries to the gneissose granite fragments marked by abrupt discordances of the gneissic foliation (Highton, 1994, fig. 3b). Inwards the breccia becomes increasingly matrix supported, with metasedimentary fragments proportionally more abundant. Rafts of metasedimentary lithologies, up to 80 m long and 20 m wide, occur towards the centre of the mass. These are elongate within the NE-trending, subvertical foliation in the quartzose amphibolite matrix. Foliated members of the post-D_3 late-Caledonian microdiorite suite crosscut the body.

Torr a' Choiltreich

Other quartzose amphibolite bodies crop out in this area. Here, two NE-trending dyke-like bodies, up to 3 m wide, crosscut the dominant S_1 foliation in the orthogneiss and amphibolites. A small lenticular mass (80 m by 20 m) [3691 0751], contains abundant xenoliths of gneissose granite, amphibolite and metasedimentary lithologies. The foliation in the quartzose amphibolite host wraps the inclusions, with quartz-feldspar segregation in the low-strain areas.

Other areas

Bodies in Coille Torr Dhùinn [3300 0600], at Bridge of Oich [3362 0370] and near Leek [3360 0402], occur as diffuse patches of hornblende-bearing gneiss and hornblende-biotite tonalite within the gneissose granite. These are considered coeval with the quartzose amphibolites. The patches are nebulous, with diffuse margins appearing to grade into the host lithology. In detail, small aggregates of biotite and amphibole mimetically replace the foliation in the granitic gneiss. Larger amphibole crystals also locally overgrow this fabric. The patches are coarse grained, leucocratic, and characterised by large (2–6 mm) crystals of amphibole and macroscopic sphene.

PETROGRAPHY

The quartzose amphibolitic rocks are medium grained, moderately mafic-rich lithologies (mafic index = 30–50). They preserve neither igneous mineralogies nor textures. A crude segregation of the mafic and quartzofeldspathic minerals into discrete layers characterises these rocks. They are petrographically distinct, comprising quartz (up to 43 modal %), with varying proportions of a blue-green potassian hastingsite, biotite and feldspar (Highton, 1994, table 1). Sphene, allanite, apatite and zircon are modally significant, with monazite and ilmenite present in trace amounts. Deformation and recrystallisation of the protolith appears to have been under amphibolite facies conditions. Orientated metamorphic amphiboles and elongate mafic aggregates, consisting of amphibole, biotite and sphene, define the foliation in these rocks.

Amphibole within the tonalitic patches (S 72198) occurs as either large poikiloblastic crystals, enclosing rounded grains of quartz, sphene and apatite, or small aggregates. Crudely aligned flakes of biotite define a weak NE-trending fabric in these rocks. Plagioclase compositions (An_{10-24}), however, are little different from the main mass. K-feldspar, biotite, allanite and rare garnet, are minor components. Contacts are gradational between the quartzose amphibolites and gneissose granite. The transition zone ranges from a few millimetres up to 0.5 m wide. Within the zone, sodic plagioclase replaces microcline, and amphibole the biotite of the gneissose granite. Grain boundaries between the feldspars are interdigitated. The new feldspar preserves relict, partially resorbed microcline.

Table 4 Representative whole-rock analyses of Fort Augustus quartzose amphibolites.

Rock type	Quartzose amphibolite					'Tonalite'
Sample no.	S 72165	S 72166	S 72167	S 72207	S 72236	S 73954
Grid ref.	347 066	347 066	373 082	349 059	349 069	328 060
SiO$_2$	58.73	65.83	59.16	67.82	69.77	68.22
TiO$_2$	2.76	2.39	2.53	1.87	2.70	0.44
Al$_2$O$_3$	6.75	7.20	9.12	8.71	5.49	15.83
FeO*	15.46	12.15	16.50	11.06	10.90	2.51
MnO	0.19	0.15	0.23	0.14	0.13	0.05
MgO	3.41	2.17	3.30	2.65	2.11	1.08
CaO	3.53	2.97	3.90	2.40	2.68	2.02
Na$_2$O	0.36	0.54	0.65	1.07	0.26	5.14
K$_2$O	3.88	3.03	3.36	3.19	2.70	2.52
P$_2$O$_5$	2.12	1.86	1.34	0.83	1.87	1.26
Loss on ignition	0.46	0.39	0.55	0.27	0.42	0.11
Total	97.65	98.68	100.67	100.11	99.04	99.78
V (ppm)	160	100	170	100	80	30
Cr	80	40	70	40	40	40
Co	28	22	30	21	21	5
Ni	35	28	30	14	22	3
Zn	226	195	185	185	176	54
Rb	300	276	224	224	275	84
Sr	32	40	63	63	43	731
Y	217	288	317	317	351	10
Zr	2056	2139	1839	2228	2653	138
Nb	49	53	41	58	59	9
Ba	700	350	660	550	390	760
La	270	400	290	520	600	20
Ce	190	100	120	350	370	10
Pb	5	6	10	15	3	17
Th	47	79	43	97	122	3
U	14	15	12	21	18	2

Total iron as FeO*

GEOCHEMISTRY

The field relationships and form of the bodies are consistent with an intrusive origin. The distinctive mineralogy and geochemistry (Table 4) are unusual in terms of known meta-igneous (or metasedimentary) lithologies in the Scottish Caledonides. The protolith and parental magma type of the quartzose amphibolite is unknown (Highton, 1994). Element mobility, because of recrystallisation during amphibolite facies metamorphism, is limited in these rocks. Thus, the systematic variations noted are compatible, to some extent, with processes recognisable in magmatic systems. Unusually low values of Al_2O_3 (< 10 wt.%) and Na_2O (< 1 wt.%), and enrichments in total iron as Fe_2O_3 (11–17 wt.%) and TiO_2 (1.8–2.8 wt.%), characterise this suite over the range of SiO_2 content (58–70 wt.%). These reflect the dominance of ferromagnesian phases during crystallisation, with little significant participation from feldspar. The enrichments in REE, high field strength (HFS) elements, Th, U, Ti and P (Figure 23), reflects partitioning into accessory phase minerals. The tonalitic patches in the granite gneiss are very poor in accessory minerals (S 73954). This is reflected in their low contents of LREE, Zr and Th. A positive correlation between Na_2O and Sr and negative correlation between Na_2O and Rb suggests biotite and K-feldspar replacement. Hybridisation within the protolith of the gneissose granite cannot explain the element enrichments and depletions. However, these apparently syn-kinematic hornblendic lithologies are similar to the unusual hornblendic rocks reported by Rock (1984) from the adjacent White Bridge area. Thus, their formation may be contemporaneous with the emplacement of the quartzose amphibolites.

LATE- TO POST-D$_3$ INTRUSIONS

Glen Moriston Vein Complex

A diffuse tracery of deformed and recrystallised granitic veins and sheets, cutting across the Moine outcrop to the south and west of Invermoriston (Figure 15), is referred to here as the Glen Moriston Vein Complex. Granitic intrusions also crosscut pre- to early Caledonian igneous rocks of the district. The complex comprises sheets and veins of muscovite-biotite microgranite, aplitic microgranite and pegmatite. Some intrusions are composite in form. Thicknesses vary from a few centimetres to a maximum of 40 m, e.g. on Binnilidh Mhor [336 157] and Càrn Dubh [3521 1210]. Most veins are less than 2 m wide. Veining is pervasive on all scales, and discordant to all structures and fabrics associated with regional D$_2$ and D$_3$ deformation in the host rocks. Most veins show some evidence of deformation (described below). Veins are largely irregular, often branching or side-stepping. The margins of the larger intrusions are curviplanar or planar, but are rarely parallel-sided. Contacts with the country rocks are sharp. There is no significant preferred orientation pattern; however, the larger sheet-like intrusions of pegmatite have a crude north to north-easterly alignment. The density of veining

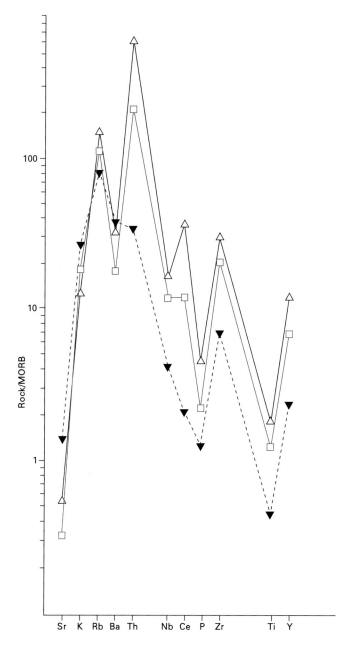

Figure 23 MORB-normalised multi-element variation diagram of the quartzose amphibolites (after Highton, 1994). Normalising factors are taken from Pearce (1983). Sample S 72236 open triangle, sample S 72165 — open square, mean West Highland Granitic Gneiss near Fort Augustus — solid triangle.

is generally less than 1 per cent of the rock volume. This increases to 5 per cent northwards and westwards from Fort Augustus towards Glen Moriston. Between Loch Tarsuinn [344 110] and Balantoul Burn [370 109], veins locally comprise 20 per cent of the outcrop. Thus the limit of veining depicted is only an approximation (Figure 2). It trends north-westwards from near Invermoriston to Loch Liath [32 19], and then along the northern flank of Glen Moriston. Both foliated and non-foliated members of the Northern Highland micro-diorite suite consistently cut across components of the Glen Moriston Vein Complex [3437 0912, 3985 0991, 4127 1387].

The microgranites and aplitic microgranites are grey-white to pinkish, non-porphyritic rocks. Grain size is generally less than 2 mm. Pegmatite intrusions are massive, medium to coarse grained (generally 10 to 50 mm), comprising large pink-white macroperthitic K-feldspar, plagioclase, quartz and muscovite, with minor biotite and garnet. There is a marked increase in grain size where the pegmatite intrusions cut pelitic or semipelitic lithologies. In these intrusions K-feldspar crystals attain 200 mm in length, with large books of muscovite. Good examples occur on Meall a' Chròm Dhoire [2790 0437], in the Allt Dail a' Chuirn [3027 0632], and in the Allt Càrn an Doire Mhòir at [3909 1228].

Composite intrusions of granitic pegmatite are found in road cuttings on the A 887 [3600 1572] and on Coire Buidhe [3511 1126]. These intrusions have thin (20 and 40 mm-thick) margins of pegmatite and cores of foliated aplitic microgranite. Contacts between these granitic phases are sharp but irregular, commonly conforming to the crystal boundaries in the pegmatite. Several intrusions of mostly coarse-grained (c. 50 mm) pegmatite are exposed in the forestry track to the south-east of Creag an Iarlam [3570 0952]. All become finer grained towards the margin. Incipient veins and apophyses of foliated biotite-microgranite are common within the pegmatitic component of these intrusions. Xenocrysts of coarse-grained quartz or feldspar are common within the microgranite. This suggests the intrusion of granitic liquids into consolidated pegmatite. An unusual type of composite intrusion, exposed within the Auchteraw Burn [3393 0973] contains discrete mica-bearing and biotite-free layers.

Evidence of late-kinematic deformation is common, with weak to well-developed foliation generally being present. This is picked out by aligned mica and elongate quartz and feldspar (S 69873, 74849). The form of individual intrusions (Plate 4) is dependant upon their original orientation. Where steeply dipping, some veins have shortened either by thickening, e.g. in the road cutting on the A 877 east of Dundreggan Dam [3600 1572]. Others have responded by folding, as in forestry track cuttings [3974 1700]. Folds are shallowly plunging, recumbent to reclined structures. Mica and elongate quartzofeldspathic aggregates define a prominent internal 'axial planar' schistosity within the veins, e.g. south of Cnoc Liath [3070 1400], and near Inchnacardoch [3897 1032]. Where shallowly inclined, the veins deform by extension. This takes the form of thinning and necking, e.g. in the Balantoul Burn [3617 1151], or boudinage, e.g. in the Allt na Criche [3894 1112]. A good example of these variations in deformational style is found in the forestry track at Creag Bheithe [3628 1438]. Here, a sheared, subhorizontal garnet-muscovite granitic-pegmatite sheet, with foliation oblique to the modified intrusive contacts, crops out. A vein of microgranite, steeply inclined, tightly folded and boudinaged, cuts this sheet.

Plate 4 Veins of granite and granitic pegmatite of the Glen Moriston Vein Complex, cutting Upper Garry Psammite Formation. Veins at low angles to bedding show evidence of thinning, while those of moderate to steep inclination are boudinaged and folded, respectfully. Road cutting on the A887 at Achlain [275 124]. (D 5072)

PETROGRAPHY

Rocks of the complex are monzogranitic (Figure 24), consisting of quartz, sodic plagioclase, microcline and biotite, with accessory muscovite, apatite, sphene, opaque minerals, garnet and allanite. They are generally hypidiomorphic to allotriomorphic-granular in texture, with evidence of some recrystallisation. Plagioclase crystals (An_{20-6}) are mostly subhedral in form. They preserve normal and oscillatory zoning (S 71692, 72224), although often overprinted by patchy zoning (S 72224, 67248). Microcline is perthitic, and together with quartz is mostly interstitial to plagioclase. Quartz, commonly strained, recrystallises to small grains or polycrystalline, often granoblastic, aggregates (S 67248, 72111). Myrmekite intergrowths are common. Biotite is brown to greenish brown in colour, mostly as anhedral aligned flakes. It is often interleaved with secondary muscovite (S 69873), and defines the tectonic fabric in these rocks. Muscovite also

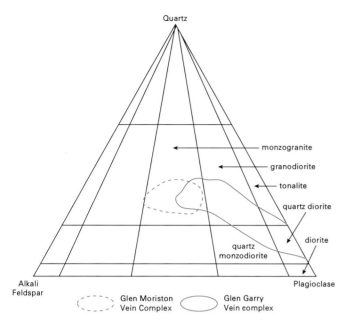

Figure 24 Modal quartz–alkali feldspar–plagioclase compositions of granitic vein complex rocks.

occurs as irregular poikiloblasts (S 67250). Accessory minerals are present in all but the coarse pegmatitic rocks, with stubby crystals of apatite, up to 0.2 mm in length, the most abundant. Sphene occurs as primary subhedral crystals, up to 0.5 mm (S 74849), and as inclusions, with zircon, in biotite. Secondary overgrowths on ilmenite are common (S 67248, 72758). Garnet, forming small (< 0.3 mm) euhedral to subhedral poikiloblasts, is present only in the microgranitic phase of the granitic pegmatite intrusions (S 70698, 72751, 72756, 72783).

GEOCHEMISTRY

Representative whole-rock analyses of the microgranite and aplitic microgranite components of the Glen Moriston Vein Complex are presented in Table 5. These show a limited compositional range (73.5–75 wt.% SiO_2), with only minor major element variations. Normative compositions within the Qtz-Ab-Or projection lie to the quartz-rich side of the cotectic (Figure 25). This suggests that these granitic rocks represent low temperature melts. The excess of Sr over Rb confirms that they are not evolved melts. The relative abundance of granitic pegmatite and pegmatitic components

Table 5 Whole-rock analyses from the Glen Moriston Vein Complex.

shows the involvement of fluids in the evolution of the suite. This may account for some scatter in the LIL data.

Syntectonic microdiorite suite

To the north-west of the Great Glen Fault, minor intrusions of calc-alkaline affinity emplaced towards the end of the Ordovician–Lower Devonian event of the Caledonian Orogeny are abundant (cf. Johnson and Dalziel, 1966; Winchester, 1976; Dearnley, 1967; Talbot, 1983). They are interpreted as forming a regional suite (Smith, 1979). The early members of the suite postdate the regional D_3 folding in the host metasedimentary rocks. However, they underwent deformation and recrystallised during the waning stage of the orogeny (Smith, 1979; Talbot, 1983). Two distinct subgroups of microdiorite and feldspar-phyric microgranodiorite are distinguishable within this early suite of the Invermoriston district, in terms of spatial distribution and mineralogy (Table 6).

METAMORPHOSED MICRODIORITES

The early components of the suite are dioritic to quartz monzodioritic in composition, typically grey to grey-green, fine grained, schistose throughout, with original magmatic textures rarely preserved. Mineral assemblages show emplacement and equilibration under epidote–amphibolite facies conditions. The intrusions form steeply dipping sheets or dykes, generally less than 1 m in thickness. Roadside exposures close to Dundreggan Dam in Glen Moriston [357 157] reveal many field relationships of this group of intrusions. Here, the microdiorite sheets cut across the regional deformational structures and fabrics in the host rocks, and also granitic veins

Sample no.	S 72757	S 72764	S 72771	S 72783
Grid ref.	3673 1449	4026 1656	3518 1547	527 1381
SiO_2 (wt.%)	74.45	74.07	73.54	74.90
TiO_2	0.03	0.22	0.28	0.15
Al_2O_3	14.60	14.21	13.89	13.55
FeO*	0.41	2.04	2.14	1.30
MnO	0.00	0.03	0.01	0.00
MgO	0.13	0.53	0.53	0.31
CaO	1.28	1.34	1.49	1.64
Na_2O	4.78	4.53	4.54	3.70
K_2O	2.43	3.46	2.76	3.52
P_2O_5	0.01	0.02	0.05	0.01
Loss on ignition	0.55	0.24	0.42	0.24
Total	98.67	100.69	99.65	99.32
Zn (ppm)	13	35	29	7
Rb	74	109	85	100
Sr	178	196	263	238
Y	9	9	18	9
Zr	28	86	194	49
Nb	1	5	8	3
La	10	20	40	10
Ba	290	650	850	740
Pb	29	26	18	33
Th	1	0	6	0
U	2	2	0	0

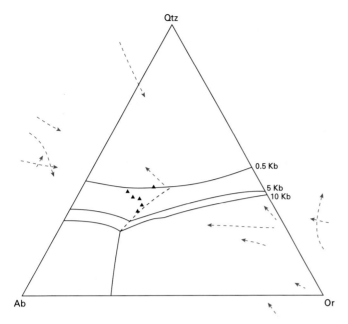

Figure 25 Quartzofeldspathic normative components of the Glen Moriston Vein Complex, projected from An onto the Qtz–Ab–Or face of the salic tetrahedron, with the trace of the eutectic at varying pressures.

of the Glen Moriston Vein Complex. Mineral assemblages comprise amphibole, biotite, plagioclase, quartz and K-feldspar, in varying proportions, with accessory sphene, apatite, zircon, allanite and opaque minerals. The mafic minerals, with quartz and, to lesser extent, feldspar, define the internal schistosity of the intrusions. This takes the form of single ribbon crystals or elongate granoblastic aggregates. Plagioclase, where fresh, is mainly oligoclase (An_{26}). Zoning is commonly absent, with crystals turbid and recrystallised to epidote. A blue-green actinolitic amphibole mimetically replaces the primary magmatic phenocrysts of mid-green hornblende. The more acidic components are weakly porphyritic, more leucocratic rocks, with biotite the predominant mafic mineral (S 72727, 72728). Both biotite and the actinolitic amphibole are oriented, defining the ubiquitous foliation. Plagioclase macrocrysts show some evidence of recrystallisation at their margins.

The intrusions mainly deformed as single incompetent layers. Internal schistosities define a sigmoidal shear fabric. The pattern of structural movement paths ('ASMOPs' of Talbot, 1982) indicate that the principal axis of strain is moderately to steeply plunging to the WSW (Figure 26). In the weakly porphyritic microdioritic intrusions, the schistosity wraps the feldspar macrocrysts. The asymmetry to the tails of quartz and feldspar, crystallised in the low strain shadows, defines a sense of overthrusting (S 70689). These kinematic indicators and rare S-C fabrics (S 70689), define a sense of shear with overturning consistently to the north-east. This sense of movement differs from ones described from the western part of the Moine outcrop (Talbot, 1983). There, deformation of the early intrusions suggests flattening shear strains ($K \approx 1$) about a moderately plunging SE-trending principal axis.

METAMORPHOSED FELDSPAR-PHYRIC MICROGRANODIORITIC SHEETS

The feldspar-phyric microgranodiorites represent a minor component of the syntectonic group of intrusions, and are restricted to the Dundreggan Forest area [31 17]. They are fine-grained, leucocratic to moderately mesocratic rocks, but are distinguishable from later porphyritic intrusions by their pervasive schistosity, and evidence of recrystallisation. They occur as E- to ENE-trending sheets, up to 3 m thick, with a shallow to moderate dip to the south.

Age relationships within this district are poorly established. However, within the Allt Rauadh [3159 1506], an unsheared mafic microdiorite dyke cuts a foliated feldspar-phyric microgranodiorite sheet. Within the adjacent Glen Affric district (Sheet 72E), similar intrusions (called early felsic porphyrites) are spatially associated with the Cluanie Granodiorite pluton (Peacock et al., 1992). Late-stage pegmatite and aplite (aplitic microgranite) veins of the pluton cut these sheets. Smith (1979) suggests that the early felsic porphyrites may be comagmatic with that pluton. To the east of the Cluanie

Rock type	Microdiorite		Feldspar-phyric microgranodiorite	
Sample no.	S 72205	S 72163	S 72725	S 71709
Grid ref.	3325 0561	3468 0656	4285 1644	3173 1490
Quartz	1.6	8.6	15.8	27.2
Plagioclase	29.0	28.7	48.3	38.3
K-feldspar	1.7	6.1	10.1	20.2
Amphibole	52.9	36.1	5.9	0.0
Biotite	0.0	16.1	11.4	14.4
Apatite	0.5	0.4	0.4	tr
Sphene	8.5	2.9	1.6	0.2
Opaques	0.0	0.8	0.1	0.1
Allanite	0.1	0.0	0.1	0.1
Zircon	0.1	tr	tr	tr
Monazite	0.0	tr	0.0	0.0
Chlorite	5.1	0.0	5.9	0.0
Epidote	0.0	0.0	tr	tr
Carbonate	0.0	0.0	0.0	0.0
Mafic index	67.2	56.3	25.4	14.8

Table 6 Representative modes of metamorphosed syntectonic microdiorite suite rocks.

Geology of the Invermoriston district

ISBN 0 11 884532 2

CORRECTION

The brown ornament for Figure 26 (p.37) was inadvertently printed on Figure 25 (p.36).
The figures should have been printed as shown below.

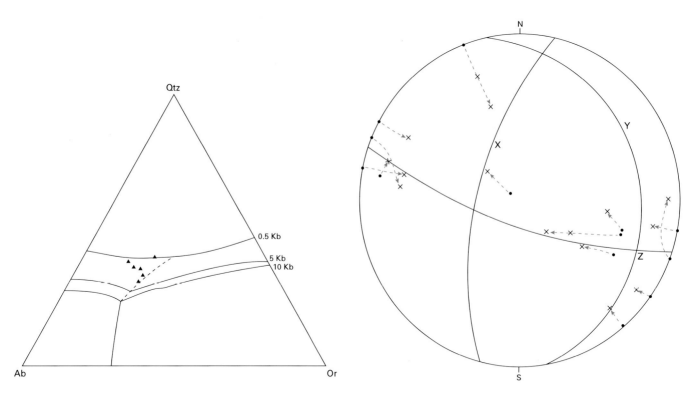

Figure 25 Quartzofeldspathic normative components of the Glen Moriston Vein Complex, projected from An onto the Qtz–Ab–Or face of the salic tetrahedron, with the trace of the eutectic at varying pressures.

Figure 26 Structural movement paths (ASMOPs) for syntectonic microdiorite suite intrusions. Solid circle — pole to intrusive contact, cross — pole to foliation element.

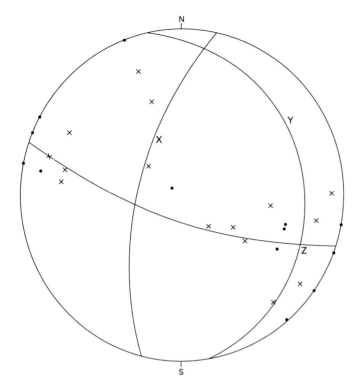

Figure 26 Structural movement paths (ASMOPs) for syntectonic microdiorite suite intrusions. Solid circle — pole to intrusive contact, cross — pole to foliation element.

mass, fragments of these porphyritic intrusions are present as clasts within the Ceannacroc breccia pipes. The intrusions of the Dundreggan Forest area clearly represent an extension of the Cluanie swarm into the Invermoriston district.

Recrystallisation or equilibration of the igneous assemblages in the porphyritic microgranodiorites suggests emplacement during the waning stages of regional metamorphism. Plagioclase dominates the macrocryst population, with subsidiary hornblende and biotite. The plagioclase macrocrysts are composite crystals, with rounded or embayed cores of andesine enclosed by rims of oscillatory zoned oligoclase-albite. Small glomerocrystic clusters may be present, but are not common. The effects of recrystallisation are variable. Where recrystallisation is weak, the feldspar becomes turbid with small grains of sericite and epidote. With further recrystallisation, albite subgrains develop at the margins of the plagioclase crystals. Ultimately, albite-quartz-epidote symplectites replace the plagioclase (S 72716). Both hornblende and biotite are preferentially oriented, defining the schistosity in these rocks. The green hornblende is largely relict, being mimetically replaced either by a blue-green actinolitic amphibole or by aggregates of decussate green-brown biotite. The microcrystalline groundmass consists of quartz, plagioclase and K-feldspar, with sphene and apatite as common accessory minerals. Elongate grains/aggregates of quartz and feldspar define a shape fabric (S 71709). This fabric wraps the plagioclase macrocrysts.

Asymmetric tails to poikiloblasts of quartz and albite are consistent with a north-easterly sense of movement.

The geochemistry of the syntectonic microdiorites is considered with that of the late- to post-tectonic microdiorite intrusions.

LATE- TO POST-TECTONIC INTRUSIONS

Glen Garry Vein Complex

Within the south-western part of the Moine outcrop, is the north-eastern part of the Glen Garry Vein Complex (Fettes and Macdonald, 1978). The complex represents a suite of apparently unmetamorphosed, late-kinematic minor intrusions of predominantly granodioritic composition, but ranges from diorite through to monzogranite. Contemporaneous with the emplacement of the vein complex are many pipe-like bodies of intrusive breccia, described separately below. The well-defined limit of the veining (Figure 2) occurs to the south of the watershed between Glen Garry, Glen Moriston and the Great Glen, south-west of Fort Augustus. The vein complex is coincident with both positive aeromagnetic and negative gravity anomalies, and is truncated by the Great Glen Fault (see Figures 36 and 37, Chapter 10). Within the adjacent districts to the south-west the magnetic anomaly is closed, with a maximum of 300 nT. The weak magnetisation of the rocks at the surface does not explain the intensity of the anomaly; however, similar perturbations are associated with other granodiorite-dominated plutons elsewhere in the Highland region, e.g. Foyers (Kneen, 1973). Thus, this anomaly adjacent to the Great Glen Fault probably reflects the presence of a plutonic mass at depth (Brown and Locke, 1979) to which the Glen Garry Vein Complex may form part of a roof zone.

The complex comprises a ramifying network of veins, sheets and larger intrusive masses. Locally these comprise up to 30 per cent of the rock volume, e.g. Allt Lundie at [292 058], but are generally less than 5 per cent. A clearly defined limit is traceable along the watershed between Glen Garry and Glen Moriston to Meall na Ruadhaig [310 087]. To the south-west of Fort Augustus, near [368 066], the line of Great Glen Fault truncates the vein complex. Only a few granitic veins or sheets extend beyond this limit, e.g. to the north and west of the Allt Phocaichain [312 102]. Most intrusions lack a preferred orientation, except within the area between Loch a' Bhainne [276 047] and Loch Lundie [296 037]. Here, there is a NNW alignment of the larger, steep-sided bodies. These intrusions are discordant, postdating the regional deformation. In sheets less than < 1 m thick, there is a pervasive but weak foliation, picked out by orientated biotite flakes. This dips at a moderate to steep angle to the south-east, e.g. west of Loch Lundie [2858 0365]. The fabric is sub-parallel to the margins of the intrusions. Boudinage, with pinch and swell structures, develop where sheets are shallowly inclined, e.g. within Coire nan Gearran [2913 0739]. The vein complex postdates the quartzose amphibolites and the Glen Moriston Vein Complex. Fettes and Macdonald (1978) report that it also postdates both foliated and non-foliated members of the microdiorite suite, although it is cut by late feldsparphyric microgranodiorites (felsic porphyrites). However, within the Creag a' Bhainne area [27 06] a small swarm of foliated leucocratic microdiorites cuts both granodioritic and granitic rocks. They in turn predate the late-stage pegmatites of the complex, e.g. in the Allt Lundie [2920 0514]. Relationships of other foliated and non-foliated microdiorite intrusions to the granitic rocks of the complex exposed within the district are unknown.

PETROGRAPHY

The granitic rocks of the Glen Garry Vein Complex are predominantly medium to coarse grained. They consist of feldspar-phyric hornblende-biotite granodiorite and biotite granite, with minor tonalite, quartz-diorite, aplitic microgranite and pegmatite (Figure 24). The larger granodiorite intrusions, e.g. Creag a' Bhainne [2720 0425], are internally homogeneous. Intrusive relationships between the compositional variants are rarely observed. However, within the Allt Lundie [2928 0427], veins of biotite granite cut across sheets of hornblende granodiorite. On Creag a' Bhainne, e.g. [2707 0402], veins and small sheets of aplite and pegmatite cut across both granodiorite and biotite granite intrusions. Textures are mostly hypidiomorphic, but becoming allotriomorphic in the more evolved biotite granites and granular in the aplitic microgranitic rocks. There is some evidence of minor recrystallisation, particularly in the prominently foliated intrusions.

Plagioclase, quartz, K-feldspar, biotite and hornblende are the main constituents, with accessory sphene, apatite, zircon and magnetite, and rare allanite and monazite. Hornblende is the principle mafic mineral in rocks of basic to intermediate composition, but is absent in the more evolved granodioritic and granitic rocks, where biotite is predominant. The habit of amphibole is dependant upon whole-rock composition. In dioritic rocks the amphibole forms large (up to 4 mm), stubby, brownish green, inclusion-free, euhedral crystals. In the granodioritic rocks the phenocrysts are commonly dark green to blue-green, prismatic crystals, with inclusions of plagioclase and sphene. In the more evolved granodioritic rocks, the amphibole is subhedral to anhedral. It occurs mainly in mafic-rich aggregates with biotite and sphene, but may be present as an interstitial mineral (S 72066). Biotite similarly varies in form, habit and abundance with rock type. Modal abundances range from 1 per cent in the diorites to a maximum of 30 per cent in some granodioritic rocks. Over the compositional range, both colour and habit vary. In the diorites, biotite forms brown, subhedral to anhedral flakes (S 73985) and aggregates (S 72207), while in the granodioritic and granitic rocks (S 70694) it forms subhedral, reddish brown phenocrysts and interstitial grains. Inclusions of zircon, apatite and sphene are common in the phenocrysts and mafic aggregates. Plagioclase is generally not a liquidus phase in the dioritic rocks, and occurs mainly within the interstices. Phenocrysts, where present, are euhedral to subhedral, lath-shaped, composite crystals. They comprise broad rims, normally zoned to oligoclase (An_{24}), enclosing rounded unzoned, commonly turbid,

andesine cores (An_{24}). The interstitial feldspar is mostly oligoclase. In the granodioritic rocks, the plagioclase forms stubby, subhedral, poikilitic, crystals up to 3 mm. Oscillatory zoned rims (An_{24-4}) enclose rounded unzoned or normally zoned cores (An_{36}). Glomerocrysts (S 72077) and synneusis clusters (S 72072) are common. Within the enclosing rims, compositional discontinuities reflect instabilities during growth. However, in many crystals the internal structures of the feldspars are often obscured by patchy zoning (S 73989). In the monzogranitic rocks (S 71435), plagioclase is allotriomorphic. Crystals show normal zoning (An_{24-8}) and are commonly antiperthitic. Alkali feldspar is a microperthitic orthoclase and generally interstitial (S 71306, 71309, 73985). Some granodioritic (S 72128) and monzogranitic rocks (S 71371, 71435) contain poikilitic macrocrysts up to 3 mm. Myrmekite is common. Quartz is interstitial in all but the monzogranitic lithologies. It occurs as either single strained crystals, with simple domain structures, or recrystallised granoblastic aggregates (S 70694, 71417). Sphene, apatite and zircon are ubiquitous in all but the more evolved granitic rocks. In some dioritic rocks (S 72066) sphene is a significant component, comprising up to 4 per cent of the mode. Apatite varies in habit from small (< 0.2 mm) euhedral, acicular crystals in the dioritic rocks to a small stubby form in the granodioritic rocks and monzogranites. Zircon is ubiquitous, forming euhedral crystals. In the granodiorites, crystals reach 0.1 mm in length. Allanite occurs only in the more mafic diorites and granodiorites, forming euhedral crystals with fine-scale oscillatory zoning (S 72077, 71337).

GEOCHEMISTRY

Fettes and Macdonald (1978) presented a petrogenetic model for the evolution of the Glen Garry Vein Complex based on fractional crystallisation of a parental quartz diorite magma under conditions of high P_{H_2O}. Their major element data is internally consistent and within the error range of the new data presented here (Table 7), except for MnO and P_2O_5 data. There is, however, some disparity in the trace element datasets. In view of the limited number of samples analysed, only broad conclusions can be drawn from variations in the datasets.

Rocks of the Glen Garry Vein Complex are predominantly metaluminous, high-K calc-alkaline, I-type granitic rocks. On Harker variation diagrams (Figure 27a) there is good correlation between most major elements and SiO_2. The dioritic and basic granodioritic rocks are weakly corundum normative. Those rocks with SiO_2 content more than 60 wt.% are diopside normative. This reflects the transition from mafic to plagioclase dominated fractionation (cf. Cawthorn et al., 1976). The trace element patterns are more complex (Figure 27b). The transition elements are most abundant in the mafic rocks; however, there is continuous depletion with increasing SiO_2 content. An almost linear depletion of V corresponds to magnetite and/or amphibole fractionation in calc-alkaline magmas (cf. Ando, 1975; Shervais, 1982). HFS elements, e.g. Zr, Y, with the LRE elements, e.g. Ce, La, show a reasonable correlation with SiO_2. The Sr and Rb content of both the quartz dioritic and tonalitic rocks is low (950 and

Table 7 Representative whole-rock analyses of rocks from the Glen Garry Vein Complex.

Rock type	Quartz diorite	Tonalite	Granodiorite	Granite
Sample no.	S 72227	S 72128	S 73985	S 73957
Grid ref.	2686 0497	2749 0386	2858 0365	2706 0402
SiO_2 (wt.%)	60.93	64.58	70.60	74.63
TiO_2	1.02	0.60	0.31	0.04
Al_2O_3	15.73	17.08	16.15	14.32
FeO*	6.03	4.21	2.10	0.52
MnO	0.07	0.04	0.04	0.01
MgO	2.42	1.24	0.49	0.05
CaO	4.35	2.70	1.88	0.26
Na_2O	4.40	5.18	4.68	4.67
K_2O	2.01	2.68	3.52	4.72
Loss on ignition	1.30	0.74	0.99	0.44
Total	98.57	99.20	100.86	99.69
V (ppm)	90	50	10	0
Cr	50	60	30	20
Co	14	10	2	0
Ni	18	7	4	0
Cu	18	11	1	0
Zn	73	65	42	5
Rb	52	80	77	132
Sr	826	811	581	91
Y	23	19	14	11
Zr	240	296	267	68
Nb	12	10	13	20
Ba	580	950	580	170
La	60	40	30	10
Ce	20	20	10	0
Pb	16	19	26	32
Th	9	7	5	1
U	4	3	2	2
CIPW norms				
Qz	12.08	14.10	23.35	28.26
Ab	38.23	44.46	39.40	39.63
Or	12.20	16.07	20.70	27.97
An	17.74	12.60	8.63	1.10
Di	1.97	0.00	0.00	0.00
Hy	14.99	10.06	5.56	1.68
Il	1.99	1.16	0.59	0.08
Ap	0.74	0.35	0.23	0.07
Cndm	0.00	1.14	1.47	1.14

50 ppm, respectively). There is a large variation from the granodioritic to monzogranitic compositions, with Sr decreasing and Rb increasing rapidly over a small SiO_2 interval. Ba is an incompatible element in the basic to intermediate rocks. However, from an inflection point at approximately 65 wt. % SiO_2 both Ba and Sr decrease. Both major and trace elements show smooth variations with increasing SiO_2. This suggests that the rocks of the complex represent a single magmatic suite. The scatter within the trace element data indicates compositional heterogeneity, through the presence of either restite (cf. Chapell and Wyborn, 1987) or fractionate accumulations of both major and accessory minerals (cf. Sawka, 1988) in metastable glomerocrystic aggregates. The distribution of most HFS and rare earth elements relates to the abundance of accessory minerals in the rocks. The correlation between Ce and Th suggests fractionation and/or accumulation of allanite, particularly in the dioritic rocks. Petrographical evidence also shows the importance of amphibole separation, particularly during the evolution of the dioritic and tonalitic rocks. Partitioning into amphibole may account for the scatter in the HFS data. The variation of Sr and Ca with SiO_2 reflects a multistage evolution of the vein complex. The diorite to tonalite sequence of rocks is little evolved in terms of the separation of a plagioclase-dominated assemblage. However, the granodiorite to monzogranite sequence can be modelled in terms of either assimilation and/or crystal fractionation (AFC) from a dioritic magma at low pressure (cf. DePaolo, 1981). The high Sr value in some granodiorites suggests either the presence of restite material or contamination. The Rb/Sr ratio in the monzogranitic rocks increases with the differentiation index, but rarely exceeds a value of 1. This points to a lack of very evolved compositions within the suite, and only minor fluid participation in the later stages.

Intrusive breccia

Many bodies of intrusive brecciated rocks occur within the outcrop of the Glen Garry Vein Complex. These occur mainly within the Moine, although two masses crop out within the Fort Augustus Granitic Gneiss. The bodies are variable in size and form. They range from dykelets less than 2 m wide to dyke-shaped bodies up to 1 km long and 200 m wide. Some irregular plugs reach 300 m in diameter. The composition of the breccia clasts reflects closely that of the host rock, with few exotic lithologies. Similar intrusive breccias occur within the Ceannacroc Lodge area [227 113] of the adjacent Glen Affric district (Sheet 72E) (Peacock et al., 1992).

Creag a' Bhainne

Three breccia bodies occur in contact with and next to the granodiorite mass outcropping on Creag a' Bhainne (Figure 28). Contacts of the two masses [2690 0409; 2737 0428], with the granodiorite are not exposed. However, the form of the western mass is transgressive. The well-exposed contacts with the enclosing metasedimentary rocks are vertical to subvertical, but with minor step-like

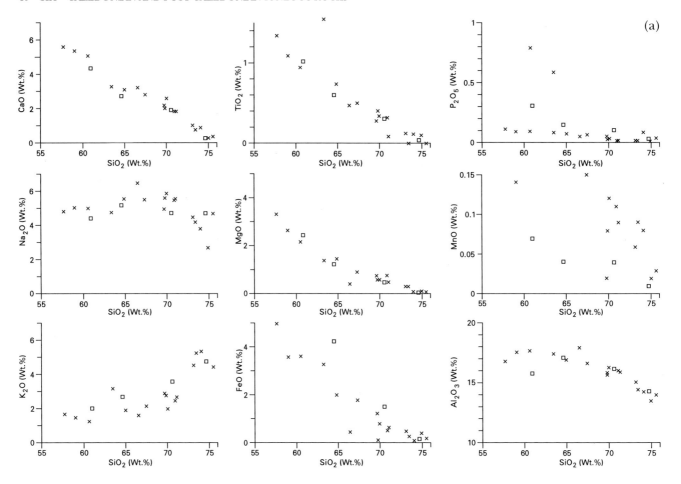

Figure 27 Harker variation diagrams for rocks of the Glen Garry Vein Complex. (a) Major oxides. (b) Selected trace elements. Open squares — new analyses by BGS; crosses — published data from Fettes and Macdonald (1978).

irregularities. By inference, the breccia bodies must postdate the granodiorites of the Glen Garry Vein Complex and the foliated microdiorites; clasts of unmetamorphosed igneous rocks are absent. The breccia clasts consist entirely of metasedimentary and/or meta-igneous country rock lithologies. The clasts are predominantly subangular to subrounded. They vary in size, but are mostly between 5 and 50 cm. Rare slab-like pieces may reach 1 m. Veins up to 4 cm wide cut vertically through the breccias. The veins are fine-grained, consisting of sand-sized grains of quartz and/or feldspar, or small rock fragments less than 0.2 mm, with a few subrounded clasts up to 5 mm.

Within the upper part of the western intrusion, fragments are mostly thin, plate-like or slabby in form, reflecting fracture mainly parallel to the principle foliation. The angularity of the fragments, which exhibit sectional ratios of between 1:5 and 1:20, is consistent with local derivation and a short entrainment within the fluidised system (cf. Platten and Money, 1987). The fragments are randomly oriented, tightly packed, with face to face contacts and no interstitial matrix (S 73962). They exhibit a greater degree of fracturing and reddening than seen in the country rocks, with evidence of displacement, dilation and rotation. This fracturing is considered to be late in

the evolution of the fluidised system, and appears largely to be a function of rotation of fragments during compaction. Clear, drusy quartz infills the porosity where dilation has occurred. This, however, does not provide a universal cement throughout the breccia mass. Within the lower 10 m of the outcrop, the packing of the clasts is more open. Here, a weakly porphyritic diorite forms the matrix. In the lowest 2 m of the exposure, diorite is predominant, with clasts in the central parts of the intrusion dispersed as 'free-swimming' xenoliths.

Lundie

A 40-m wide, north-west-trending breccia dyke or elliptical plug occurs to the north-west of Lundie [2910 0463]. The intrusion is composite, with an outer matrix-absent rim breccia enclosing an inner core of xenolithic quartz diorite. The breccia consists of angular to subangular clasts, predominantly of metasedimentary lithologies, with minor schistose amphibolite and pegmatite. Contacts with the host rocks are nowhere seen. However, in the Allt Lundie [2923 0445], there is a pronounced reddening of the country rocks within 1.5 m of the first exposure of the breccia. A steeply inclined, fine-scale (< 3 cm) block fracture is pervasive throughout this zone. This is orien-

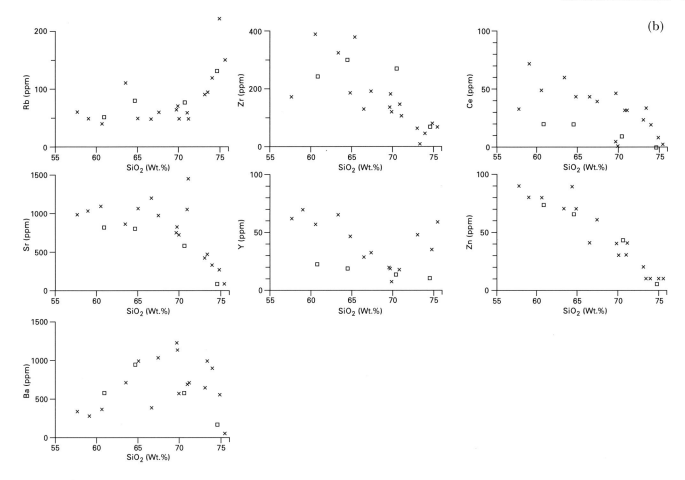

(b)

Figure 27   (*Continued*)

tated subparallel to the trend of the breccia mass. A ramifying network of thin dark red veins and veinlets, up to 1 cm wide, cut the country rocks of this marginal zone to the body. These are aphanitic to microcrystalline, but often contain rounded grains of quartz and feldspar, and small rounded rock fragments (S 71315). A drusy quartz cement is present in several fractures, showing dilation within the external fracture system coeval with the emplacement of the breccia mass. The rim breccias have a bimodal clast population which consists of massive irregular blocks up to 2 m in length interspersed with smaller clasts. The reddening and closely spaced fracture of the latter are similar to that of the wall-rocks. Thus, incorporation was through spalling from the wall-rock into the intrusion. Clast shapes are little modified and packing is tight but with an interstitial cement of sand-sized material and drusy quartz. The north-western part of the mass, e.g. at [2909 0465], contains angular fragments between 0.01 m and 2 m across. Fragments of psammite, amphibolite and pegmatite are tightly packed. Finer-grained material, drusy quartz and, more rarely, microdiorite infill the poorly sorted clast-supported framework. The clasts show little evidence of abrasion. Their size and shape reflect closely the lithology, with interbanded psammite and semipelite or semipelitic rocks forming large blocky fragments. Psammite gives rise to blocky and/or slabby clasts with length to thickness ratio as high as 10:1. Bleaching of the outer layer (< 2 cm) of these large well-rounded rock fragments is common.

The inner part of the Lundie intrusion consists of a xenolith-rich, medium-grained quartz diorite. The contact of the inner diorite with the breccia is sharp, with no evidence of chilling and only limited filling of the porosity. Metasedimentary xenoliths are mostly subrounded with embayed, corroded and diffuse margins. Thin (< 2 cm), pale coloured, feldspar-rich rims enclose the fragments. These rims are gradational into a pale grey, hornblendic granite and then into the quartz diorite. The hornblendic granite (S 71411) is essentially a hybrid rock. Small (0.2–4 cm), incipient pools of graphic granite and mafic-rich schlieren are abundant. These patches of granophyre probably represent partial melts of the metasedimentary rocks.

Torr a' Chait

A large breccia mass, only part of which crops out within the Invermoriston district, lies to the west of Loch Lundie on Torr a' Chait [2880 0322]. The NNW-trending dyke-

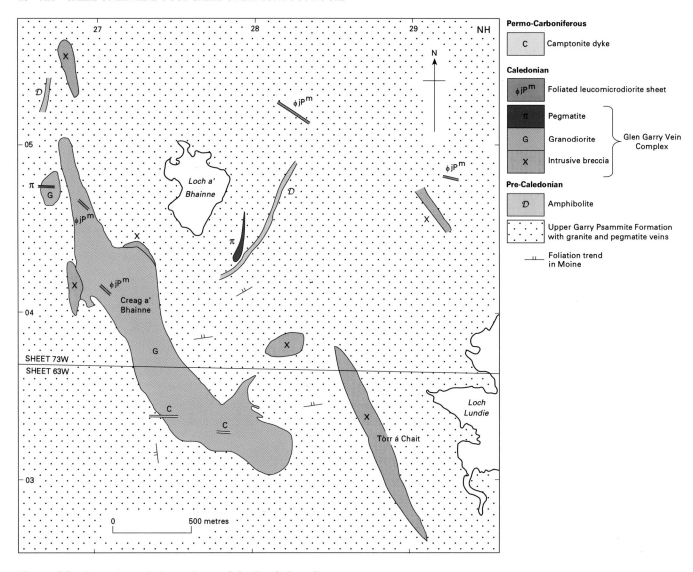

Figure 28   Late-tectonic intrusions of the Loch Lundie area.

like body, up to 1300 m long and up to 100 m wide, differs in containing a weakly foliated, xenolithic granodiorite matrix, e.g. at [2860 0363]. Angular fragments of psammite, with some schistose amphibolite and, more rarely, gneissose granite, dominate the clast population. In the matrix-absent breccia [2884 0313], clasts range from sand-sized particles of 3 mm to blocks mostly less than 0.3 m. Rafts of psammite up to 15 m in length and 3 m wide are common in the granodiorite. The margin of the body, where exposed, is gradational with stepped contacts. A zone of reddening and intense fracturing between 2 and 5 m wide encloses the intrusion. This becomes progressively disrupted, with a 0.2 m-wide transition zone of partially rotated blocks that passes into the marginal breccia. Within the host psammite, fracturing along both the lithological banding and the foliation controls the step-like form of the contacts. Spalling of these foliation/joint-controlled blocks, produces a well-defined angular, faceted contact. Aplitic microgranite veins cut both the breccia and granodiorite.

Other bodies

Other significant intrusive breccia pipes crop out to the north of Creag a' Bhainne [268 055], and to the north and east of Meall Mor [328 045]. Contacts are nowhere exposed; however, both are plug-like in form, with a north to south elongation. Reddening of the host rocks is common, with a drusy quartz cement infilling dilational joint fractures. Angular to subangular blocks, up to 2 m in diameter, of gneissose granite are predominant in the Meall Mor mass. Metasedimentary and metabasic rocks are minor components in this mass. In both bodies, leucocratic microdiorite partially permeates the breccia porosity. This forms either a localised ramifying network in the breccia or isolated blind pockets. Deformation and recrystallisation of the microdiorite matrix are evident, with biotite flakes and aggregates defining the schistose fabric (S 71411, 72093). This fabric appears to predate consolidation of the breccia, and infilling of the remaining porosity by drusy cements.

The fabric may wrap around perturbations in the clast surface, with quartz deposited within the low-strain areas (S 72083). The body to the north of Creag a' Bhainne illustrates well the process of consolidation in these intrusions. Here, quartz and/or chlorite fills the step fractures in the plate-like fragments. Slickensides develop on the hanging wall of the fracture surfaces. Where plate-like clasts are at a high angle to underlying fragments, buckling of the upper plate is common. Fracturing appears to supersede the initial plastic flow as buckling increases. Fractures developed on the upper surfaces of the clasts are infilled by quartz. The intense fracturing in parts of the intrusion suggests that reduction of the pre-cement porosity during consolidation was by an order of magnitude. This is mainly through an increase in packing density.

A minor pipe-like body crops out to the west of Bridge of Oich [333 037]. Other NW-trending dyke-like masses, occur on Meall na Ruadhaig [316 085] and in the Allt Phocaichain [318 104].

Microdiorite suite and appinite

Most of the minor intrusions recorded within the district, both to the north-west and south-east of the Great Glen Fault, are of calc-alkaline affinity. They range in composition from basic to acid-intermediate. The field and petrological classification of the components of the suite follow, wherever possible, the IUGS recommendations (Le Maitre, 1989). The boundary between microdiorite (sensu stricto) and mafic microdiorite is at M (mafic index) = 50, with the prefix 'leuco' assigned to those rocks with M< 25. Typically subgroups within the suite form a continuum with transitions from one petrographical type to the next. The appinitic rocks have affinities with groups within the microdiorite suite.

The microdiorite suite intrusions, emplaced mainly as easterly dipping parallel-sided sheets, are generally undeformed. Talbot (1982; 1983) interpreted the development and intensity of the internal shear fabrics as a function of thickness and orientation in the regional stress system, while Smith (1979) invoked a west-directed overthrusting of Moine Thrust age. There is little documentation of the geochemistry of the suite, or their role in the plutonic evolution of the Northern Highland terrane. Their age relations to the other intrusive groups within the district are rarely observed.

Appinite

Appinite intrusions are not common within the district. Their distribution coincides with an ENE-trending belt of microdiorite suite intrusions through Glen Moriston. The appinite intrusions form either elongate masses, e.g. at Beinn Bhreac [3016 1628], or subcircular plugs, e.g. north of Dalchreichart [2915 1415] and [2885 1430].

A good example of the plugs forms a prominent feature on the hillside to the west of Binnilidh Beag [3192 1545]. This concentrically zoned intrusion is approximately 40 m in diameter. It comprises coarse-grained rocks that range from hornblende-pyroxene-bearing

ultramafite at the outer margin (S 71696), through phlogopite-ultramafite (S 71697) and mesocratic appinitic diorite (S 71694), to a core of feldspathic appinite (S 71695). Internal contacts are gradational over a few tens of millimetres to several metres. Oikocrysts of clinopyroxene, up to 50 mm across, dominate the outer facies, with large phenocrysts of green-brown hornblende, green-brown biotite, apatite and sphene. Carbonate and microcline mainly infill the interstices, with subordinate plagioclase and quartz. The carbonate may be magmatic, poikilitically enclosing phenocrysts of biotite, apatite and sphene. Secondary late-magmatic crystallisation is extensive, with the primary assemblages overgrown by coarse-grained plumose aggregates of actinolite. Aggregates of green-brown biotite mimetically replace hornblende. The phlogopite-ultramafite is texturally and mineralogically heterogeneous. It varies from rocks in which phlogopite, as colour-zoned crystals up to 20 mm across, is the only mafic mineral in the rock, to phlogopite-hornblende or phlogopite-actinolitic amphibole-bearing rocks. Sphene and apatite are common accessory minerals, with feldspar generally absent. In the mesocratic appinitic diorite, euhedral crystals of green-brown hornblende, up to 50 mm long, contain cores infilled by quartz and albite, and enclose early formed apatite and sphene. Aggregates of green-brown biotite pseudomorph the hornblende. Small crystals of exsolved ilmenite occur at the rims of the pseudomorphs. These are set in a matrix of coarse-grained albite and subsidiary K-feldspar, with minor plumose aggregates of oligoclase.

A pink leucocratic rock forms the core of the body. This facies contains euhedral to subhedral mesocrysts of orthoclase up to 10 mm across. These poikilitically enclose euhedral crystals of hornblende and biotite and, less commonly, small macrocrysts of oligoclase. Abundant plumose and spherulitic clusters of albite, up to 4 mm in diameter, form the predominant texture in this rock. The spherules, generally, have nucleated on the macrocrysts. The interstices are infilled by carbonate or microgranular aggregates of granitic composition. Pyrite is abundant.

Microdiorite suite

Most of the minor intrusions within the district form part of this suite. Although present throughout the district, their main distribution is in an ENE-trending belt through the Glen Moriston area. They range in composition from basic–intermediate (in which four compositional groups, A to D, are recognised) to granodioritic. Orientation data reveals two swarms, one predominantly NNE- and the other east-trending. The NNE swarm consists of Group A and D intrusions, and the east-trending swarm Groups B and C. Components of the latter swarm are consanguineous with intrusions cutting the Cluanie Granodiorite pluton. Most intrusions are moderate to steeply dipping, parallel-sided sheets ranging up to 7 m thick. They preserve chilled margins against the country rocks, and igneous textures and mineralogies. However, there is some evidence for emplacement under conditions of waning regional metamorphism. Flow fabrics within the chill zones are picked out by the crude alignment of phenocrysts, with minor irregularities in the wall-rocks giving rise to swirl structures. In most intrusions, shear fabrics are absent; where recognisable, the shear fabrics are consistent with a subhorizontal east–west shortening.

PETROGRAPHY

In the field three sets of rock type are recognisable: basic and intermediate to acid microdiorites (*sensu lato*) and microgranodiorite. The most basic set is divisible into four distinct groupings based on their P and Ti chemistry, and together form a continuum from mafic microdiorite to microdiorite. Representative modes are presented in Table 8.

Basic–intermediate microdiorite

Group A: High P/High Ti   Rocks of this group are typically grey-green, medium to moderately coarse grained and melanocratic (M=65–80), commonly with feldspathic net-vein segregation. Disequilibrium assemblages are common, with many primary textures and mineralogies overprinted by incipient late- to post-magmatic crystallisation. They typically contain dark green-brown hornblende primocrysts, generally enclosed by large interstitial plagioclase or less commonly K-feldspar. Quartz and biotite are also interstitial, with rare pools of aplitic microgranite. This latter texture suggests melt separation, which gives rise to the macroscopic net-veining in these rocks. Clinopyroxene occurs only as rare, resorbed cores within hornblende (S 72723). Plagioclase is mostly andesine, weakly zoned to oligoclase. Feldspar phenocrysts are uncommon, although radial glomerocrystic aggregates and spherules are present (S 72723). Apatite and sphene are modally abundant (up to 4%), with minor allanite, zircon and ilmenite. Minor late- to post-magmatic recrystallisation is recognisable. A blue-green actinolitic amphibole epi-

taxially overgrows hornblende in a reaction with plagioclase. Less commonly, aggregates of green-brown biotite pseudomorph the hornblende. Plagioclase may become turbid and replaced by sericitic mica and epidote.

Group B: High P/Low Ti   Intrusions of this group are mineralogically quite distinct, and the more numerous of the high P groups. Most are grey to grey-green, fine to medium grained, slightly more leucocratic rocks (M = 40–60). Net-veining is absent. Amphibole phenocrysts are fusiform, up to 2 mm long. Zoning is common, from brown pargasitic hornblende cores to rims of mid-green hornblende. However, there is partial or complete replacement by a blue-green actinolitic amphibole. Biotite is present only as rare overgrowths on the hornblende (S 71733). The interstices comprise plagioclase with minor quartz and K-feldspar. The rocks are texturally heterogeneous, with zoned (An_{32-8}) poikilitic plates of plagioclase subordinate to radial, plumose and spherulitic aggregates of fusiform crystals (S 72769). The feldspars forming the spherules are mostly less than 0.5 mm in length (but 2 mm is not uncommon). Zoning is weak within these crystals (An_{28-25}). The intraspherulite space is mostly infilled with quartz, and more rarely microgranite. From experimental work, similar growth forms develop only with large rates of undercooling (Kirkpatrick, 1975; Swanson, 1977). Apatite, as acicular crystals up to 0.5 mm, is the principal accessory mineral, with subordinate zircon and ilmenite. Sphene is mostly secondary, replacing the ilmenite. Epidote, where present, is an interstitial magmatic mineral, poikilitically enclosing apatite, zircon and allanite.

Group C: Low P/High Ti   These predominantly melanocratic rocks (M = 45–80) are of basic–intermediate composition. They are petrographically similar to rocks of Group A, although apatite is rare. Amphibole primocrysts are of mid to light brown pargasitic hornblende, characterised by cores infilled with quartz and albite, and

Table 8   Representative modes of samples from the late to post-tectonic microdioritic suite.

Rock type	Basic-Intermediate Group								Acid-intermediate microdiorite			Feldsparphyric microgranodiorite	
	High P/High Ti		High P/Low Ti		Low P/High Ti		Low P/Low Ti						
Sample no.	S 72723	S 71735	S 71733	S 72769	S 72753	S 71732	S 72762	S 72741	S 72786	S 72747	S 72109	S 71335	S 72775
Grid ref.	2734 1346	2736 1396	2731 1333	3328 1501	3653 1461	2728 1317	4010 1664	4072 1699	3585 1571	3860 1582	3275 0751	2920 0514	3610 1380
Quartz	0.2	1.2	1.3	9.0	3.2	9.2	1.9	6.3	6.4	13.6	13.8	15.6	25.3
Plag.	17.7	25.0	31.1	53.5	22.0	40.4	20.4	36.6	70.3	62.8	61.5	55.7	37.5
K-feldspar	0.2	0.4	0.9	3.5	0.7	2.0	0.3	2.1	3.1	3.8	8.2	7.7	22.1
Amphibole	32.0	34.0	59.4	42.0	53.1	33.6	63.9	47.3	13.6	10.7	7.6	2.0	0.0
Biotite	39.0	32.9	0.6	0.5	17.4	12.1	11.0	1.4	0.6	1.3	7.3	20.0	13.0
Apatite	4.2	2.0	3.2	1.3	0.6	0.5	1.0	0.8	0.7	0.7	0.2	tr	0.3
Sphene	4.0	2.3	0.9	0.8	0.1	0.2	0.9	1.0	2.2	2.1	0.6	tr	1.2
Opaque	0.0	0.1	tr	0.2	2.4	1.5	0.4	0.2	0.5	0.4	tr	0.1	0.2
Allanite	0.5	0.2	0.4	tr	tr	tr	tr	0.7	tr	tr	0.1	tr	0.1
Zircon	0.4	tr	0.5	tr	tr	tr	tr	0.3	0.1	tr	0.1	0.0	0.0
Monazite	0.1	tr	0.1	tr	tr	tr	tr	tr	0.0	tr	tr	0.0	0.0
Chlorite	0.0	0.0	0.0	0.0	0.0	0.0	0.0	0.0	0.0	0.0	0.0	0.0	0.0
Epidote	tr	tr	tr	tr	tr	tr	0.3	1.1	2.5	4.3	1.1	0.0	0.0
Carbonate	0.0	1.4	1.2	0.0	0.0	0.5	0.0	1.7	0.0	0.0	0.0	0.0	0.0
Mafic index	80.0	70.0	63.0	43.5	73.0	47.0	77.0	52.0	14.1	15.0	16.0	22.0	15.0

exsolved ilmenite (S 72753). Pale green actinolitic amphibole pseudomorphs, or irregular aggregates of actinolite and a pale brown biotite, replace the hornblende. Plagioclase is predominantly interstitial, although tabular primocrysts are present in some rocks. These are mostly turbid, with patches enclosing irregular crystallites of actinolite and biotite. Ilmenite and titano-magnetite are the most predominant accessory minerals with allanite, zircon and monazite. Sphene is secondary, replacing the opaque phases.

Group D: Low P/Low Ti This group of intrusions is the predominant component of the NNE-trending swarm. They are green-grey, medium-grained, melanocratic rocks (M=70–80), ranging from mafic microdiorite to quartz diorite. The acicular phenocrysts of hornblende, zoned from mid-brown cores to brownish green rims, define a prominent magmatic foliation. In rocks of basic composition, plagioclase forms weakly zoned (andesine An_{35-25}) interstitial poikilitic plates. At more acid compositions (52–55 wt.% SiO_2), the large feldspars poikilitically enclose small subhedral laths of labradorite (An_{65-60}). Quartz and K-feldspar are subordinate and interstitial. Accessory minerals are similar to those of Group C rocks, with sphene secondary after ilmenite.

Late-magmatic replacement may be extensive with, apart from sphene, actinolite and biotite replacing hornblende. Biotite may be locally abundant, forming large anhedral flakes or aggregates that overprint the magmatic fabric (S 72763). The replacement of hornblende by the actinolitic amphibole also involves plagioclase. Epiphytic zones of turbid plagioclase enclosing dendrites of actinolite are present at the interface between the feldspar and secondary amphibole (S 72762).

Intermediate to acid microdiorite

Rocks within this group lie in the range 56 to 66 wt.% SiO_2 and form a continuum from quartz microdiorite to microgranodiorite ('microdiorite' and 'leucomicrodiorite' of Smith, 1979). They are fine to medium grained, grey to buff-grey, mesocratic rocks (M=15–30), consisting essentially of plagioclase and hornblende, with biotite and subordinate interstitial quartz and K-feldspar. Many intrusions have little modified magmatic textures, and are panidiomorphic. Two distinct plagioclase populations are present in these rocks (S 72786). Euhedral oscillatory zoned macrocrysts of andesine (An_{45-34}), up to 3 mm, commonly form glomerocrystic clusters (S 72747). The macrocrysts are enclosed by a finer-grained groundmass consisting mainly subhedral, normal and/or oscillatory zoned plagioclase crystals, of oligoclase-andesine (An_{42-15}) composition. Irregular narrow rims of albite (An_{6-10}) enclose both plagioclase forms. Phenocrysts of hornblende are fusiform, up to 2 mm in length and zoned from brown cores to green-brown rims. Ilmenite exsolution in the amphibole is common, while replacement by a pale green actinolitic amphibole and pale brown biotite is not extensive. At more acid compositions, phenocrysts of dark brown biotite are predominant. The accessory minerals include euhedral apatite, ilmenite, zircon, zoned allanite and

sphene. The interstices are infilled by granular aggregates of granodioritic composition, with subordinate biotite, green hornblende and sphene.

Feldspar-phyric microgranodiorite (felsic porphyrite)

An ENE-trending belt of these intrusions extends from Loch Cluanie in the Glen Affric district (Peacock et al., 1992, fig. 38), eastwards into Glen Moriston. There it tapers out to the south of Loch a' Chràthaich [37 19]. Other minor swarms are found in the Glen Coiltie [46 27] and the Auchteraw Wood [33 07] area. These intrusions are sheets with an average thickness of 5 m, but range up to 15 m. All have a shallow to moderate dip to the south-east.

The rocks are porphyritic, with macrocrysts predominantly of oscillatory zoned oligoclase and subordinate biotite, hornblende, sphene, apatite, allanite, zircon, and ilmenite. They grade, both texturally and compositionally, into the intermediate–acid group (S 69878). Large macrocrysts of plagioclase, up to 6 mm across, contain resorbed cores of calcic andesine enclosed by narrow irregular rims of albite (S 71736). Fusiform, brown hornblende phenocrysts, up to 2 mm, are abundant, with a dark brown biotite subordinate. At compositions > 66 wt.% SiO_2, hornblende is absent. The macrocrysts lie within a fine-grained to aphanitic groundmass, which ranges from granodioritic to granitic in composition, with a granular texture (S 72704). The incoming of the two-fold plagioclase macrocryst population marks the transition into the intermediate-acid group of rocks. As in all other components of the suite, evidence of late-magmatic/low-grade autometamorphic recrystallisation is common. The secondary mineral assemblage consists of blue-green actinolitic amphibole, green-brown biotite, albite, epidote, quartz, sphene and carbonate.

GEOCHEMISTRY OF THE SYNTECTONIC AND LATE- TO POST-TECTONIC MICRODIORITE SUITES

The end-Caledonian suites of minor intrusions within the district, whether of syntectonic or post-tectonic age, are of similar compositional range (Tables 9–11; Figures 29–30). Both show high-K calc-alkaline affinity, and typical calc-alkaline evolution trends, although the appinitic group shows minor iron-enrichment (Figure 29). For most of the syntectonic, late- to post-tectonic microdiorite and feldspar-phyric microgranodiorite suites major element distributions are indistinguishable (Figure 30a). All show an inverse relationship between SiO_2 and FeO*, MgO, CaO, MnO, P_2O_5 and TiO_2. Variation in P_2O_5 and TiO_2 contents is not significant within the low P and low Ti groups until approximately 60 wt.% SiO_2; at this point there is a marked inflection. A similar lack of variation in P_2O_5 is found in the data from the syntectonic microdiorite suite, while TiO_2 decreases over the compositional range. The high P/high Ti group of intrusions, with some components of the syntectonic suite, have elevated K_2O content similar to the appinitic rocks. In the post-tectonic groups both Al_2O_3 and Na_2O increase with increasing SiO_2 to the inflection point, beyond which these elements remain relatively constant. Similar variations are found in the syntectonic suite,

Table 9 Whole-rock analyses of representative samples from the syntectonic microdiorite suite.

Rock type	Metamorphosed microdiorite				Metamorphosed feldsparphyric microgranodiorite	
Sample no.	S 72200	S 72205	S 70689	S 72778	S 72784	S 71710
Grid ref.	3336 0561	3325 0561	3570 1572	3570 1572	2853 1353	3159 1506
SiO_2 (wt.%)	50.50	52.19	60.10	64.06	66.11	67.83
TiO_2	1.14	1.05	0.99	0.65	0.38	0.36
Al_2O_3	13.79	14.11	16.30	16.27	15.60	16.10
FeO*	10.49	9.15	6.47	4.44	2.98	2.65
MnO	0.18	0.16	0.12	0.04	0.01	0.03
MgO	8.01	8.37	3.22	1.63	1.09	1.13
CaO	6.94	7.56	4.43	3.61	2.91	2.14
Na_2O	0.08	3.04	5.23	5.44	6.46	6.56
K_2O	3.30	2.11	1.49	1.80	1.84	2.30
P_2O_5	0.18	0.21	0.27	0.23	0.11	0.11
LOI	3.49	1.80	0.75	0.57	2.03	0.68
Total	98.10	98.75	99.37	98.74	99.52	99.90
V (ppm)	170	120	100	50	40	40
Cr	350	180	40	40	40	30
Co	32	25	17	8	7	6
Ni	39	137	44	9	6	6
Cu	41	40	—	6	1	5
Zn	69	68	76	70	66	68
Rb	134	43	50	29	39	42
Sr	448	601	876	920	811	899
Y	18	16	18	12	8	7
Zr	131	190	187	229	161	160
Nb	8	6	—	11	7	6
Ba	350	400	540	720	610	680
La	40	60	70	50	50	40
Ce	20	20	—	30	20	20
Pb	11	15	17	16	20	27
Th	2	3	3	5	5	4
U	2	2	—	2	1	3

although the bivariate behaviour of Al_2O_3 is less distinct (Figure 29).

The minor intrusions, in general, are crystal–liquid hybrids. In-situ fractionation is unlikely to have been of great significance within intrusions of this size. Therefore, patterns within the variation diagrams are likely to be composite trends. These broadly reflect the evolution of a single composite or multiple source at any given time. The linear variation with increasing SiO_2 suggests some similarity in source composition, with the decrease of Fe, Mg and Ca indicating fractionation of pyroxene, amphibole and/or biotite. Concomitant increase of Al, Na and K in rocks of basic–intermediate composition suggests only minor feldspar fractionation until the inflection point. The variation of both P and Ti suggests the removal of apatite, sphene and/or ilmenite as liquidus phases. This may involve either a multiple or a single but replenished source. High field strength (HFS) elements, e.g. Zr, Nb, Th and U, preferentially partition into phases such as zircon and monazite. Their behaviour in the late- to post-tectonic groups is bivariate (Figure 30b). These elements are low in abundance within the low P groups and syntectonic suite intrusions. They have a flat to slightly positive trend over the range of SiO_2 content. The high P groups show an enrichment of HFS elements. This may represent the presence of zircon as a restite mineral entrained within the melt, or unusually higher Zr solubility in the parental liquid. The light rare-earth elements (LREE) are similarly bivariate in their distribution. This probably reflects an abundance of acceptor phases, such as allanite and sphene, in the high P group rocks.

The large ion lithophile (LIL) elements, e.g. Rb, Ba and Sr, follow only in part the trends defined by K and Ca. All high P groups show an enrichment in Ba and Sr and are depleted in Rb as compared with both the low P groups and syntectonic suite rocks. Feldspar fractionation cannot account for these patterns, as confirmed by the flat distribution pattern of Y. Similar enrichment/depletion trends are found in the appinitic rocks. However, the heterogeneity of the appinites reflects the volatile-rich nature of their magmas. This may lead to ambiguity in any interpretation of the element distribution patterns. In rocks of more acid composition (> 60 wt.% SiO_2), the trends of the LIL elements and Y are indicative of feldspar fractionation.

PRE-CARBONIFEROUS IGNEOUS ROCKS SOUTH OF THE GREAT GLEN FAULT

The igneous rocks described in the previous sections all crop out to the north-west of the Great Glen Fault. A similar, end-Caledonian (late-Silurian to lower Devonian) calc-alkaline intrusive history is found within the Central Highlands. Part of this terrane occurs within the Invermoriston district to the south-east of the Great Glen Fault. Direct comparison is, however, not possible. Plutonic activity is post-tectonic, represented here by the emplacement of the Foyers Plutonic Complex (Marston, 1971) at about 410 Ma (Pankhurst, 1979). Emplacement of the pluton was at mid to upper-crustal levels (Highton, 1986). Only a small part of the complex occurs within the district. Minor intrusions are uncommon in this part of the district, and are represented mostly by felsite sheets (which are not found in the area to the north of the Great Glen Fault) and microdiorite sheets.

Table 10 Whole-rock analyses from the Binnilidh Beag appinite plug.

Sample no.	S 71694	S 71696	S 71697	S 71695
Grid ref.	3192 1545			
SiO$_2$ (wt.%)	38.71	40.80	41.10	59.10
TiO$_2$	1.73	1.24	1.19	0.30
Al$_2$O$_3$	12.14	7.57	9.03	18.50
FeO*	15.02	7.44	7.89	3.46
MnO	0.27	0.13	0.15	0.05
MgO	5.94	12.01	10.60	0.54
CaO	11.02	17.60	14.00	4.38
Na$_2$O	2.77	1.41	2.14	8.24
K$_2$O	3.46	2.51	8.30	1.47
P$_2$O$_5$	2.02	2.57	1.61	0.05
H$_2$O+	5.15	2.32	2.01	1.04
CO$_2$	0.16	4.81	6.50	2.18
Total	98.39	100.41	99.52	98.27
V (ppm)	240	100	110	40
Cr	60	50	420	50
Co	33	29	28	3
Ni	11	111	132	2
Cu	161	18	38	126
Zn	229	67	102	25
Rb	38	43	69	24
Sr	2988	1237	1619	4355
Y	43	24	28	39
Zr	498	210	301	856
Nb	30	9	20	17
Ba	7210	1400	1510	2840
La	740	290	590	1320
Ce	280	80	260	860
Pb	32	8	16	51
Th	34	9	47	105
U	6	1	5	13

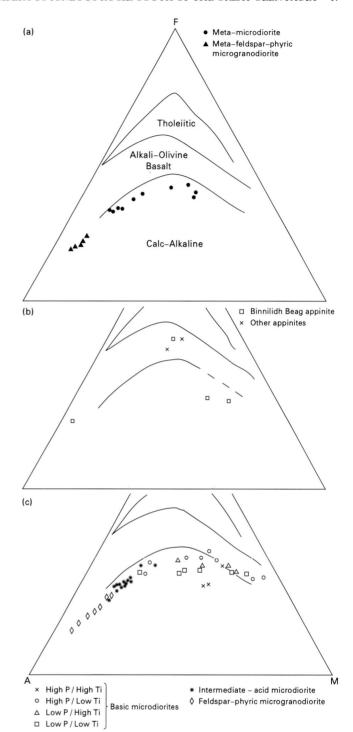

Figure 29 MgO–FeO*–total alkalies discriminant diagrams for: (a) early (syntectonic) microdiorite suite intrusions; (b) appinites; (c) late to post-tectonic microdiorite suite intrusions. Fields from Irvine and Barager (1971). FeO* = total Fe as FeO.

Foyers Plutonic Complex

On the evidence of gravity and aeromagnetic data, plutonic rocks of the Foyers Complex underlie much of the Loch Knockie area [45 13] at shallow depth. The outcrops in this area do not represent satellite intrusions, but the partial unroofing of the complex. Extensive veining and disruption of the carapace of metasedimentary rocks by granitic rocks is directly relatable to the pluton (Marston, 1971). The veins and satellite bodies comprise grey, essentially medium- to coarse-grained porphyritic granodiorite. The granodiorite consists of plagioclase, microperthitic orthoclase, green hornblende, biotite and quartz, with accessory sphene, apatite, allanite and zircon. Some secondary epidote and chlorite may be present. Plagioclase phenocrysts form antiperthitic subhedra with oscillatory zoning from andesine cores to oligoclase rims, commonly enclosed by irregular overgrowths of albite. K-feldspar is mostly interstitial with quartz and minor plagioclase, but also forms large macrocrysts. The macrocrysts, up to 20 mm in length, are commonly poikilitic. They enclose all other phases except quartz.

A poorly exposed area of appinitic diorite outcrops within the Loch Knockie area [449 137]. This is a large

xenolithic mass enclosed by the granodiorite. The appinites are medium- to coarse-grained rocks. They consist of large prominent stubby crystals of dark green amphibole set in a heterogeneous, predominantly

Table 11   Whole-rock analyses of representative samples from the late to post-tectonic microdiorite suite.

Rock type	Basic-intermediate microdiorite								Acid-intermediate		Feldsparphyric microgranodiorite	
	High P/High Ti		High P/Low Ti		Low P/High Ti		Low P/Low Ti					
Sample no.	S 72723	S 71708	S 71733	S 72769	S 72753	S 71732	S 72762	S 72741	S 72768	S 70688	S 71704	S 72770
Grid ref.	2734 1346	3173 1490	2731 1333	3328 1501	3653 1461	2728 1317	4010 1664	4072 1699	3205 1534	3570 1572	3016 1787	2754 1468
SiO_2 (wt.%)	43.08	51.86	44.93	57.57	47.32	54.01	49.01	54.63	60.88	64.60	65.09	69.13
TiO_2	1.68	1.30	1.05	1.12	1.60	1.14	1.07	1.00	0.94	0.81	0.58	0.37
Al_2O_3	9.88	12.33	8.23	16.50	12.64	15.43	11.10	14.37	16.82	16.10	16.16	16.29
FeO^*	11.47	8.00	8.23	6.84	10.55	8.30	10.02	7.43	5.78	4.75	3.62	2.16
MnO	0.17	0.12	0.17	0.08	0.16	0.13	0.19	0.10	0.06	0.08	0.03	0.00
MgO	12.46	10.12	14.73	3.14	11.98	5.53	12.80	6.34	2.69	2.10	1.65	0.77
CaO	10.61	6.57	12.46	5.51	8.13	6.56	8.74	6.16	3.66	3.94	3.00	2.19
Na_2O	1.74	2.84	1.98	5.65	2.69	4.13	2.34	4.46	5.17	5.26	5.64	5.44
K_2O	2.94	3.98	0.38	1.06	1.01	1.72	1.54	1.50	1.78	2.25	2.27	2.93
P_2O_5	1.11	0.52	0.71	0.42	0.20	0.27	0.21	0.25	0.35	0.29	0.22	0.12
LOI	4.11	2.04	6.21	2.07	2.33	1.19	2.06	2.07	1.77	0.97	0.94	0.43
Total	99.25	99.68	99.98	99.96	98.61	98.41	99.08	98.31	99.90	101.15	99.20	99.83
V (ppm)	160	130	230	120	230	150	180	120	90	70	50	20
Cr	170	900	390	60	550	110	890	180	40	20	40	30
Co	39	42	32	22	46	25	44	25	13	20	9	4
Ni	343	394	306	22	264	65	289	137	13	—	10	2
Cu	7	7	59	11	32	15	20	40	10	2	7	2
Zn	142	99	83	120	68	85	87	68	79	73	61	51
Rb	41	9	76	52	29	34	46	43	51	51	39	55
Sr	1582	786	1046	1126	489	699	321	601	1138	1094	1282	1037
Y	20	17	14	155	20	19	17	16	17	14	10	3
Zr	335	174	339	278	141	174	123	190	247	230	207	169
Nb	21	16	23	10	5	8	6	6	8	—	2	0
Ba	1520	1070	1910	690	260	750	240	400	870	960	870	970
La	290	150	230	110	40	70	70	60	90	80	70	60
Ce	100	50	80	60	20	20	30	20	50	—	40	30
Pb	11	8	11	8	10	18	6	15	14	24	23	22
Th	11	7	8	9	1	4	2	2	7	6	5	5
U	4	3	6	1	1	1	2	2	1	1	1	1

feldspar-rich groundmass. Marston (1971) reports the presence of relict cores of clinopyroxene in these amphiboles.

A thermal metamorphic aureole is more restricted in its development on the south-western side of the Foyers pluton (Tyler, 1981) compared with that to the north-east (Highton, 1986). Rocks within the roof zone are horn-felsed. Here, there is a loss of regional metamorphic fabrics and mineralogies, and extensive growth of both sillimanite and K-feldspar. More commonly the assemblage cordierite + andalusite + fibrolitic sillimanite is found. The emplacement level of the pluton was between 12 and 14 km (3.9–4.1 kb), with maximum aureole temperatures of 650–680°C (Tyler and Ashworth, 1983; Highton, 1986).

Minor intrusions

FELSITE

Felsites are the more predominant minor intrusions to the south-east of the Great Glen Fault Zone. They form small stockworks of sheets in the Loch nan Eun [45 09] and Lochan Màm-chuil [43 11, 44 11] areas, with no preferred orientation. These felsite intrusions are similar to those in the Foyers district (Sheet 73E). They are granitic in composition and spatially associated with a granite–granodiorite vein complex that predates the emplacement of the Foyers pluton (Highton, 1986). Contacts are sharp, mostly crosscutting, with variably developed chilled margins. Internal fabrics are rare, and restricted to textural heterogeneities represented by the presence of coarser-grained domains. The felsites comprise (S 74432) an essentially allotriomorphic to granular groundmass of quartz and feldspar, although patches of graphic intergrowth are common. Mafic minerals are absent. Xenocrystic aggregates of oscillatory zoned plagioclase and quartz may be present, with euhedral to subhedral macrocrysts of microcline and quartz. Alteration of the groundmass feldspar to sericitic mica and haematite is often extensive.

Salmon-pink-coloured, fine-grained to aphanitic, weakly to non-porphyritic intrusions, which cut granodioritic rocks within the Lochan a' Choin Uire area [460 164], represent late-stage aplitic microgranite differentiates of the Foyers Plutonic Complex.

MICRODIORITE

Microdiorite intrusions are uncommon in the south-east part of the district, with only two small steeply inclined sheets recorded on Creag Coire Doe [430 066] and on the western flanks of Càrn Clach na Feàrna [423 089]. At outcrop they are grey to green-grey weathering with prominent phenocrysts of amphibole. Net-vein segregations of leucogranodiorite are common. Contacts, where exposed, are sharp and cross-cutting with well-developed chilled margins up to 20 mm thick. A weak flow fabric is picked out at the marginal zones by aligned spindles of amphibole. In thin section (S 74430,31), the rocks are similar to the microdiorite suite exposed to the north-east of the Foyers pluton. These have been shown to be comagmatic with the plutonic complex (Highton, 1986). They comprise fusiform, subhedral crystals of green-brown hornblende, mostly replaced by a pale green actinolitic amphibole, set in a groundmass of poikilitic andesine with minor interstitial quartz and rare K-feldspar. The plagioclase plates pass into pools of fusiform crystals as radial, plumose and spherulitic aggregates. Individual crystals are subhedral, mostly less than 0.3 mm in length (but 0.5 mm is not uncommon), weakly zoned andesine (An_{35-39}). Accessory minerals include sphene, apatite, zircon and pyrite. Secondary alteration is extensive with abundant irregular grains and aggregates of epidote replacing the feldspar.

PERMO-CARBONIFEROUS DYKES

Intrusions of Permo-Carboniferous age are not numerous in the district, and are restricted to the outcrop north-west of the Great Glen Fault Zone. All are camptonites, and represent peripheral members of the easterly trending Eil-Arkaig swarm (Rock, 1983). Intrusions within the district mostly coincide with this trend, although some trend more north-easterly. These trends coincide with the post-Devonian fracture pattern. Two of the large dykes, in the Allt Dail a' Chùirn [3075 0593] and in the Allt na Graidhe [3170 0780] show evidence of crushing at their margins, with small developments of haematite–carbonate-rich breccia, up to 0.2 m wide. Although the intrusions are mostly parallel-sided dykes, they may locally be sinuous or bifurcating. They range in thickness from 0.5 to 10 m, but are mostly less than 1.5 m. At outcrop, the rocks are typically dark grey-green to black. They are aphanitic, but often contain macroscopic phenocrysts of olivine and clinopyroxene, with small amygdales (< 3 mm). The latter preferentially weather out, to produce a characteristic mottled, 'pock-marked' surface. The larger dykes contain a margin-parallel compositional lamination. These comprise milli-metre-scale accumulations of mafic phenocrysts, occurring up to 0.3 m from the external contacts (S 71382, 71386). Other internal variations include laminar structures, reflecting variations in grain size (S 70278), and veining of the porphyritic variants by aphyric material (S 71384).

PETROGRAPHY

The camptonite dykes are aphanitic to fine grained, and predominantly melanocratic. Most are porphyritic and locally contain up to 60 per cent phenocrysts (S 70277, 70278). The macrocryst population consists mainly olivine and clinopyroxene, with rare biotite; less commonly there are glomeroporphyritic aggregates of pyroxene (S 71383). The pyroxene phenocrysts are euhedral to subhedral acicular crystals up to 2 mm, but mostly less than 0.3 mm. All display a pronounced colour zoning (S 71382, 71383) from colourless/pale green cores (?diopsidic augite) to pale brown/purple-brown rims (?titanaugite). Replacement by secondary minerals is variable, with the development of mimetic aggregates of actinolite and chlorite, or carbonate in some rocks.

Figure 30
Harker variation
diagrams of
syntectonic and
late to post-
tectonic
microdiorite
suite intrusions.
Symbols as
Figure 29.
(a) Major
oxides.
(b) Selected
trace elements.

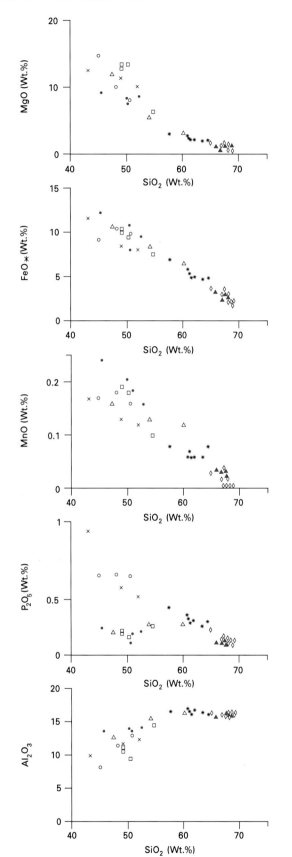

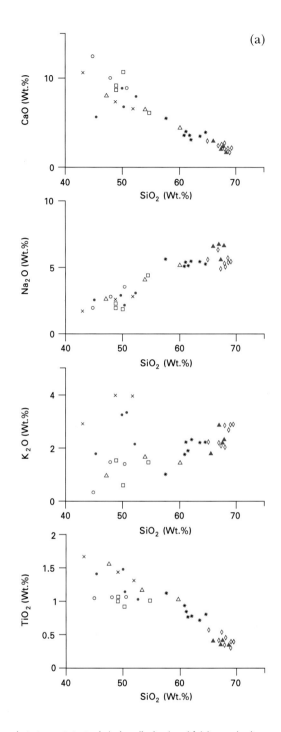

(a)

Late to post- tectonic 'microdiorites' and felsic porphyrites
× High P/High Ti   ✶ Intermediate–Acid
○ High P/Low Ti   ◇ Feldsparphyric microgranodiorite
△ Low P/High Ti
□ Low P/Low Ti

Syntectonic 'microdiorites' and 'felsic porphyrites'
● Meta-microdiorite
▲ Meta-feldspar–phyric microgranodiorite

(Symbols common to Figures a and b)

(b)

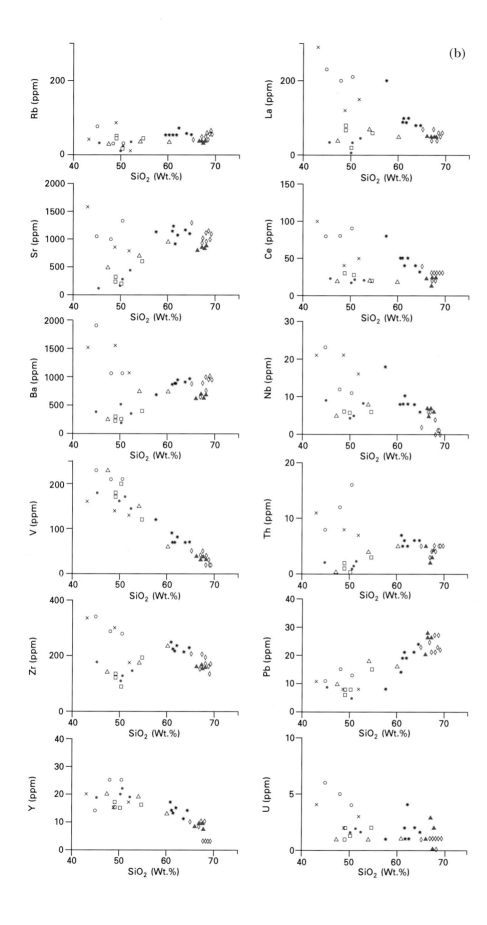

Olivine is abundant in these rocks, locally consisting up to 50 per cent of the rock (S 70277). Both macrocrysts, up to 2 mm in length (S 70278), and microphenocrysts, of approximately 0.2–0.5 mm, are common. They are rarely fresh (S 70278), commonly pseudomorphed by serpentine, chlorite, smectite, opaque minerals and carbonate (S 70277, 71386, 72201). A dark red-brown amphibole, presumed to be kaersutite (cf. Rock, 1983), is everywhere microphenocrystic. Small idiomorphic crystals of a dark red-brown biotite are rare in the camptonites (S 72201).

The groundmass is mostly fine grained (< 0.2 mm), consisting of feldspar, pyroxene, amphibole, biotite and nepheline, with accessory apatite, magnetite, ilmenite, spinel, and carbonate. The groundmass is not resolvable where aphanitic (S 70277). In one sample (S 70278), poikilitic plagioclase encloses the mafic and accessory minerals. Felsic minerals are predominant in the groundmass (50–70 per cent), with plagioclase the most abundant and with nepheline a minor constituent (< 5 per cent). Plagioclase alters to a mixture of carbonate and sericitic mica in many cases. Where fresh, compositions range from An_{30} to An_{46}. Nepheline alters to a pale green, coarsely crystalline mineral aggregate of 'nepheline-x' (cf. Rock, 1983), which encloses relicts of the feldspathoid. The mafic minerals largely reflect the phenocrystic phases, with titanaugite, amphibole and biotite, but olivine is absent. The amphibole is generally predominant over pyroxene, and biotite is a minor constituent although in sample (S 72202) biotite is in excess of the amphibole.

Feldspar in the groundmass exhibits a wide range of morphologies both within and between intrusions. In the coarser-grained rocks the feldspar is commonly acicular, forming an intergranular framework to the other phases (S 71384). Dendritic crystal forms (cf. Lofgren, 1974; Lofgren and Gooley, 1977) are predominant, varying from parallel grain clusters (S 72094, 71383) to fan (S 72095) and plumose structures (S 71384). Feldspar ocelli, a texture common in the Permo-Carboniferous camptonite suite (Rock, 1977; 1983), form prominent dendritic patches or spherulites. Co-existing pyroxene and amphibole may be intergrown in a spherulitic texture with feldspar. Single or multiphase amygdales of carbonate, chlorite or chalcedony are common. They have concentric growth forms, with outer rims of chalcedony or chlorite, enclosing cores of spherulitic chlorite (S 72095), carbonate and/or chalcedony (S 70278, 71382).

GEOCHEMISTRY

Analyses presented in Table 12 are of the margin and centre of a 1.7 m dyke from the Invervigar Burn [3334 0562]. The intrusion is typical of the camptonitic alkali lamprophyres from the Eil-Arkaig swarm (cf. Rock, 1991), with basanitic compositions and norms. Variations from the edge to the centre of the intrusion are minor. Only K, Rb, Sr and Ba show any significant increase, while Ca, Cr and Ce decrease.

Table 12 Whole-rock analyses of a Permo-Carboniferous camptonite dyke.

Sample no.	S 72201	S 72202
Grid ref.	3345 0562	
	margin	centre
SiO_2 (wt.%)	45.67	46.86
TiO_2	0.07	0.94
Al_2O_3	15.38	15.88
FeO*	8.39	8.33
MnO	0.15	0.15
MgO	6.40	6.94
CaO	8.81	6.87
Na_2O	3.01	2.85
K_2O	3.14	4.17
P_2O_5	0.61	0.67
Loss on ignition	7.31	5.39
Total	99.84	99.05
V (ppm)	140	130
Cr	120	90
Co	26	27
Ni	84	85
Cu	37	35
Zn	61	60
Rb	129	218
Sr	985	1016
Zr	151	156
Y	17	18
Nb	14	17
La	110	120
Ce	50	40
Ba	1430	1840
Pb	10	7
Th	3	3
U	2	1

SEVEN

Lower Devonian (Old Red Sandstone)

The unconformable base of the Devonian succession does not crop out in the Invermoriston district. All contacts with the metamorphic basement rocks are tectonic. Conglomerate, breccia and arkosic sandstone occur in a fault-bounded strip south-west of Fort Augustus which extends into the Glen Roy district (Sheet 63W). Exposure is poor, and the rock is shattered due to the proximity of major faults.

The well-exposed area of Devonian strata around Mealfuarvonie [45 22] forms part of a large outlier which has been described by Mykura and Owens (1983). The outcrop is up to 3 km wide and consists of a faulted, tilted and locally thrust sequence at least 1800 m thick. This consists of fine-grained, predominantly red and largely planar-bedded sandstones, interbedded with partly fault- and thrust-bounded lenses of breccio-conglomerate and conglomerate (Figures 31, 32).

North of Drumnadrochit in the adjacent Foyers district (Sheet 73E) a number of pale purple and pale grey mudstone beds have yielded poorly preserved plants and a rich microflora which suggests an Emsian or possibly lower Eifelian age. No fish remains have been found and the sediments are ascribed to the topmost Lower Devonian, partly on the evidence of the microspores and partly by comparison with similar sediments in northern Scotland.

The greater part of the north-western junction of the Old Red Sandstone with the Moine is a steeply inclined fault, but there are short stretches where the junction is gently inclined. At Loch a' Bhealaich [450 208] the short east–west-trending section of the junction is nearly horizontal and overlain by steeply inclined breccio-conglomerate and sandstone, suggesting that in this sector the junction is a thrust.

The breccio-conglomerates of the outlier were probably deposited in a series of alluvial fans at the foot of fault scarps. The fine sandstones are the deposits on the piedmont plain which may have occupied a valley along the Great Glen Fault Zone.

Because of the tectonic complexity of the district, and the lack of any definite outcrop of the basal unconformity, it has not been possible to establish a meaningful stratigraphical succession for the entire outcrop. The lithological description of the sedimentary rocks is therefore on a regional basis, commencing with the breccio-conglomerates, followed in turn by the pebbly grits and the sandstones.

BRECCIO-CONGLOMERATE

Area west of Alltsigh [44 19]

The sedimentary rocks in this area are medium to fine conglomerates intercalated with arkosic pebbly grits. The conglomerates consist of rounded clasts up to 200 mm in diameter, set in a matrix of coarse sandstone composed mainly of fresh angular feldspar grains. A high proportion of the pebbles are of leucocratic non-foliated biotite-granite; metamorphic rocks of Moine type are rare. As there are no outcrops of post-tectonic granite within the adjoining Moine terrane on the north-west side of the Great Glen Fault, the source of the granites must have lain on the south-east of the fault.

Creag Dhearg [456 205]

The breccio-conglomerate forming the ridge of Creag Dhearg has a maximum thickness of 320 m and thins out, by intercalation with pebbly sandstone, towards the north-east. It is poorly graded, with individual clasts ranging in size from 2 to 60 cm. A high proportion of the large clasts are subangular. They consist of Moine psammite with subordinate gneissose granite, microgranite, pegmatite, aplite and vein quartz. Clasts of non-foliated post-tectonic granite are virtually absent, but the matrix is arkosic with a high proportion of fresh feldspar clasts.

The breccio-conglomerate is underlain by pink fine- to medium-grained planar-bedded sandstone with lenses of pebbly grit and gritty arkosic sandstone. It is overlain by a variable thickness of gritty arkosic sandstone with pebbly lenses and, on the north-eastern slopes of Creag Dhearg, the conglomerate passes by interdigitation into pebbly arkose.

In the thickest part of the breccio-conglomerate outcrop, just west of the summit of Creag Dhearg, the bedding is chaotic with little or no alignment of clasts and no trace of imbrication, suggesting a debris-flow origin.

Loch a' Bhealaich [449 207]

Breccio-conglomerate crops out above the thrust plane immediately north of Loch a' Bhealaich. It is up to 200 m thick at the thrust but passes north-eastwards, by intercalation, into arkosic sandstone within a distance of 500 m. The breccio-conglomerate is coarse in the south-west but becomes progressively finer towards the north-east. It is similar in clast and matrix composition and in texture to the breccio-conglomerate of Creag Dhearg.

Mealfuarvonie [458 223]

Conglomerate and breccio-conglomerate forms a NE-trending outcrop which is up to 3.5 km long and between 600 and 700 m wide. The conglomerate may be up to 400 m thick near the summit of Mealfuarvonie, where its base is locally exposed along the south-east slope of the hill. The south-eastern margin of the conglomerate is in part a fault or thrust and in part a normal sedimentary junction (Figure 31). South-east of the summit [461 219] there is a transitional zone up to 50 m

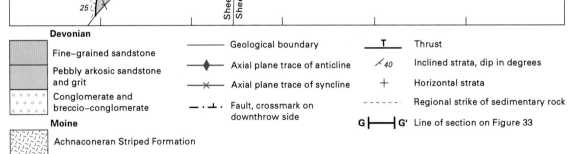

Figure 31   Geological map of the Lower Devonian (Old Red Sandstone) rocks west of Loch Ness.

thick, between the underlying medium-grained sandstone and the coarse unbedded conglomerate. This zone contains medium conglomerate with ribs and lenses of red-brown pebbly sandstone.

At one locality 1.15 km north-east of the summit of Mealfuarvonie [467 227], the massive conglomerate lenses out abruptly towards the south-east exposing a steeply inclined eastward-dipping upper junction and a near-horizontal lower junction. This suggests that the south-eastern margin (toe) of the original gravel fan lay close to the south-eastern boundary of the present conglomerate outcrop. The north-western margin of the Mealfuarvonie conglomerate is everywhere faulted. North-eastwards the conglomerate lenses out by interdigitation with sandstone.

In the southern and central parts of the outcrop the conglomerate is massive, virtually unbedded and devoid of sandstone intercalations. Individual beds of conglomerate

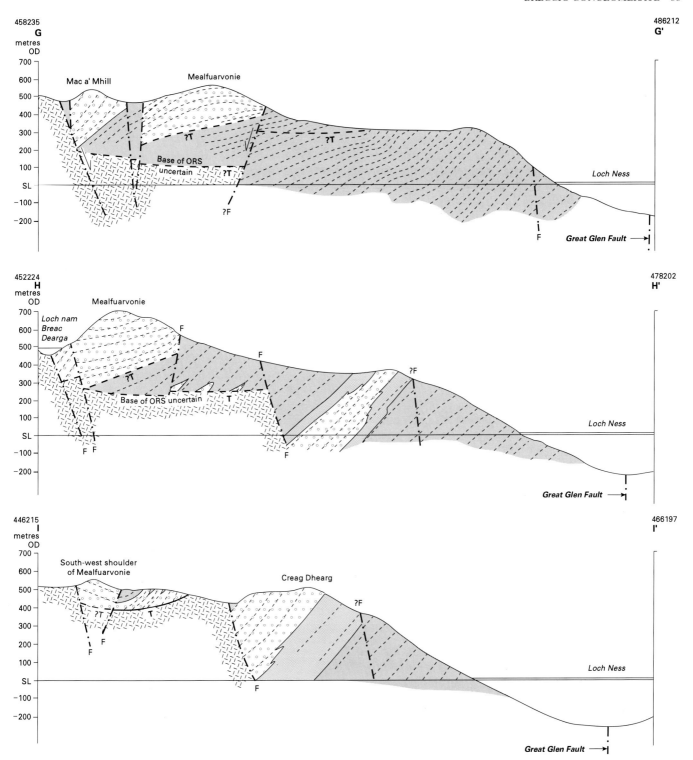

Figure 32 Cross sections through the Lower Devonian (Old Red Sandstone) outcrop west of Loch Ness (G–G' to I–I' on Figure 31). Key as Figure 31.

cannot usually be distinguished. It is poorly graded and consists of subangular to subrounded clasts ranging up to 300 cm, but with some scattered larger clasts up to 600 mm in diameter. The clasts show no preferred orientation. Many are matrix-supported and consist predominantly of Moine psammite and subordinate semipelites, pegmatites, aplites and some gneissose granite. The matrix is a coarse purplish brown gritty sand, with a relatively low proportion of feldspar grains. A high proportion of these conglomerates are debris flow deposits.

On the northern and north-eastern slopes of Mealfuar-vonie the conglomerate is generally finer and contains thin lenses and intercalations of purple pebbly sandstone, generally not exceeding 100 mm in thickness. The clasts in the conglomerate here tend to have their long axes parallel to the bedding, but there is little evidence of imbrication. The matrix is purple, gritty, with quartz grains predominating over feldspar. As in the area to the south-west, psammite clasts predominate, forming 95 per cent of the total. Other clasts include pegmatite and rare microgranite. The conglomerate lenses out north-east-wards into the Foyers district (Sheet 73E) by inter-digitation with sandstone.

It is probable that the eastern and north-eastern margins of the Mealfuarvonie conglomerate approximate roughly to the original eastern margin of the alluvial fan. There is also some south-westward interdigitation between conglomerate and sandstone at the base of the Mealfuar-vonie conglomerate, suggesting that the margins of the fan to the south and south-west were also not far from the present outcrop. This limited evidence would suggest that the source of the Mealfuarvonie conglomerate lay to the west and that the fan was located along a north to NE-trending fault scarp close to Loch nam Breac Dearga.

Nighean a' Mhill [458 231] and Mac a' Mhill [460 234]

Conglomerate forms the two small, but prominent hills, Nighean a' Mhill and Mic a' Mhill, north-east of Loch nam Breac Dearga. Both hills are bounded by faults (Figure 32) and the dip of the sedimentary rocks within the fault blocks ranges from 40 to 60° to the west. The strike of the strata within the blocks diverges from the regional strike of the Mealfuarvonie rocks by up to 90°. This suggests that the fault blocks were rotated at some stage in their tectonic history.

The Mac a' Mhil conglomerate rests on red sandstone and reaches a thickness of 240 m. The lowest 15 m consist of fine conglomerate interbedded with medium-grained sandstone and pebbly sandstone. This is overlain by massive, poorly bedded conglomerate with thin, widely spaced partings of sandstone. In the lower 30 m of the conglomerate pebbles do not exceed 15 cm in diameter, but the overall clast size increases upwards. In the upper part of the sequence some blocks reach 0.5 m. Clasts are generally subrounded, flattened Moine psammites with subordinate granite and gneissose granite. They are closely packed with an interstitial sandy matrix. The conglomerate of Nighean a' Mhil is similar to that of Mac a' Mhil and is probably part of the same fan deposit. It passes upwards via fine conglomerate with thin sandstone ribs into purple, fine-grained, planar-bedded sandstone with thin siltstone partings, exposed on the shore of Loch nam Breac Dearga.

PEBBLY ARKOSIC GRIT

The breccio-conglomerates of Alltsigh, Creag Dhearg and Loch a' Bhealaich are underlain, overlain and inter-digitated with pink medium- and coarse-grained pebbly arkosic sandstones. Arkosic sandstones with scattered pebbles, locally with thin lenses of conglomerate, are also interbedded with predominantly fine-grained sandstones on the hill slope north and north-east of Alltsigh.

The pebbly arkoses around Alltsigh are immature and appear to have been derived directly from a local source of leucogranite. Individual grains are angular to sub-angular, poorly graded, within the medium to coarse sand range, and consist of up to 70 per cent feldspar. The matrix is in places composed entirely of carbonate, though a fine siliceous matrix is developed locally.

Coarse-grained arkosic grit with scattered pebbles, generally up to 50 mm in diameter, is interbedded with purple fine-grained sandstone within 3.5 km north-east of Alltsigh. Planar bedding is dominant, though some small-scale cross-bedding has been recorded. Graded beds and upward fining cycles are rare. The proportion of feldspar grains decreases from 70 per cent at Alltsigh to 40 per cent some 2 km further north-east, although this is not a uniform reduction.

In the ground north-east of Loch a' Bhealaich, between the breccio-conglomerates of Creag Dhearg and Mealfuar-vonie, pebbly arkosic sandstone with conglomeratic lenses extends about 1 km north-eastwards from the conglomerate outcrops shown on Figure 32 and interfingers with pink fine- to medium-grained sandstone with small-scale cross-bedding. The proportion of feldspar grains in the pebbly grits is between 50 and 70 per cent, which suggests a nearby granite or, more probably, gneissose granite source area. The pebbles are, however, mainly of quartzite, psammite and gneissose granite, suggesting that the rivers contributing to the alluvial fans of this area had traversed a varied metamorphic terrain.

SANDSTONE

Most of the sedimentary rocks within the area are purple, fine- to medium-grained micaceous sandstones, generally planar-bedded, but poorly laminated. Large-scale cross-bedding is rare and individual cross-bedded sets are generally less than 50 cm thick. Ripple-marked surfaces are common, as are thin partings of deep purple mudstone between the sandstone units. Many mudstones have desiccation cracks, and rip-up clasts of purple mudstone are commonly incorporated in the basal few centimetres of the overlying sandstone sets. Small-scale disturbed bedding, slump and water-expulsion structures have been recorded throughout this sequence, but are rare. There are few fining-upward cycles of the type produced by high-sinuosity streams, though many of the thin purple mudstone and siltstone beds may represent over-bank deposits. A proportion of the sandstone beds have load-cast and, more rarely, flute-cast bases and some show rudimentary grading. This suggests that some sandstone beds may have been laid down by turbidity currents in playa lakes. Most sandstones, however, show no grading or bottom structures, and they were probably deposited by ephemeral streams on the piedmont plain beyond the toes of the alluvial fans.

EIGHT

Faults

Numerous faults occur throughout the Invermoriston district, those associated with the Great Glen being the most important (Figure 33). The main faults referred to are labelled A to G on this figure.

Great Glen Fault Zone

The Great Glen Fault Zone divides the Invermoriston district into two parts. It operated as a transcurrent structure during the later stages of the Caledonian Orogeny and was subsequently re-activated, not only during the Devonian, but also during later periods. The direction, scale and timing of movements remain the subject of continuing research, but the net displacement is sinistral and may

exceed 100 km (Johnstone and Mykura 1989). The net vertical displacement across the Great Glen Fault Zone is not known. The structure of the Great Glen is not a single fault plane but a zone, up to 1.5 km wide, bounded by the Great Glen and Glen Buck faults (A and B) of unknown but probably large displacement. The Aberchalder Fault (C) separates Old Red Sandstone from psammite. The psammite is intensely microfractured and partly converted to cataclasite. The cohesive cataclasite is cut by numerous steeply inclined crush zones infilled with soft breccia and clay gouge. It is likely that the development of cohesive cataclasite predates the deposition of the Old Red Sandstone. This has been affected by later movements only, associated with widespread shattering and the production of fault gouge. The Glen Buck Pebbly Psammite Formation to the south-east of the Glen Buck Fault is little affected by breakage. Outcrops of Fort Augustus Granitic Gneiss north-west of the Great Glen Fault are also only slightly affected. However, on Torr a' Choiltreach [369 075] and Creagan a' Chip [365 070] the gneiss is cut by thin, red to purple irregular veinlets of cataclasite, up to 3 cm wide. The gneissose granite adjacent to the Caledonian Canal at [3640 0742] is brecciated. Both the Moine and gneissose granite are cut by ramifying networks of hydrothermal alteration veins, often prehnite-bearing, within a 2 km-wide zone adjacent to the fault.

Faults around Mealfuarvonie

The phase of thrusting and faulting of the Old Red Sandstone around Mealfuarvonie (Figures 31, 33) is of late or post-Lower Devonian age and closely connected with the lateral and vertical displacements along the Great Glen Fault Zone. South-west of Alltsigh the Old Red Sandstone is thrust over and faulted against the Moine (Figures 31, 32). The thrust plane is inclined at approximately 25° to the south-east. The Old Red Sandstone close to the thrust is shattered and cut by many small faults but the rock fabric is not greatly changed. The Moine rocks, by contrast, are intensely brecciated within a belt, up to 170 m wide at outcrop, beneath the thrust. The rock in this zone was originally psammite and is now converted to cataclasite net-veined by calcite.

The most intensely crushed rock is a cataclasite (S 62447–8), consisting of irregular 'eyes' of quartzo-feldspathic granofels up to 5 mm long, set in a dark grey to black aphanitic matrix. Within it some small crushed fragments of quartz and feldspar are still recognisable. The proportion of clasts within the rock can be as low as 25 per cent. The matrix is cut by a network of irregular veinlets of calcite which also forms diffuse patches.

Most other samples from the crushed zone are proto-cataclasite or crush breccia (S 62451–2, S 61559). The

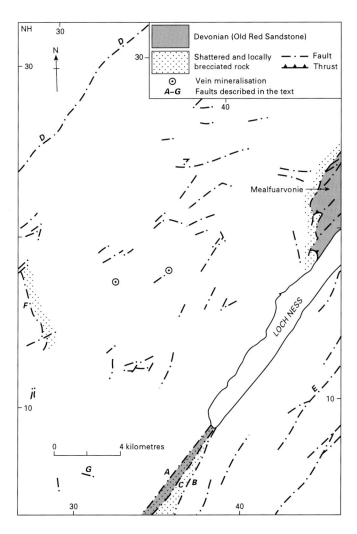

Figure 33 Faults and mineralisation in the Invermoriston district.

clasts range up to 5 cm in size and the matrix forms a network around the clasts. In areas where the larger clasts are widely spaced, the crushed matrix is fairly coarse. Where they are closely packed then the matrix is usually very fine and almost completely isotropic. Irregular anastomosing calcite veinlets are always present.

The plane of the major fault bounding the Mealfuarvonie outlier in the north-west is nowhere exposed. Its relationship to the topography of the area suggests that it is steeply inclined to the south-east. This corresponds with the inclination of the parallel, more southerly, boundary fault which extends from Loch a' Bhealaich southwards to Meall na Sròine. The latter is well exposed on the north side of the Allt Saigh gorge [451 196], where it is inclined at 70° to east. Here, a 0.3 to 1.2 m-wide crush zone is bounded to the west by a 30 m-wide zone of cataclased Moine metasedimentary rocks. These faults truncate the thrusts and they may have been formed during a period of relief of pressure after the phase of NW–SE compression.

At Nighean a' Mhill and Mac a' Mhill a number of east- to NE-trending faults branch off from the western boundary fault. The Old Red Sandstone enclosed by these faults is not only steeply inclined but its strike has been rotated by nearly 90° relative to the regional strike. The conglomerate of Meall a' Choire, enclosed by two faults, is similarly rotated. A possible explanation for the anomalous strike and dip of these conglomerates is that they represent parts of thrust sheets, originally occupying a different tectonic level to that of the country rock to the south-east. They were juxtaposed subsequently by normal faulting.

Strathglass Fault

Erosion along the line of the Strathglass Fault (D) has given rise to the broad valley of Strathglass. As with so many large-scale faults in the Scottish Highlands, no indication of the magnitude of displacement can be obtained from the local geology. The main evidence of the proximity of the fault takes the form of fracturing and reddening of rocks in stream sections perpendicular to, and on both sides of, Strathglass. The intensity of the fracturing and reddening varies considerably from section to section. However, the intensity seems to be greater on the south-east side of the valley.

Stratherrick–Loch Mhor Fault

The Stratherrick–Loch Mhor Fault (E) is a NE-trending vertical fault that has been recognised in Gleann nan Eun [448 098] and traced north-eastwards to the eastern margin of the district. The disposition of the lithostratigraphical units from outside the Invermoriston district implies a sinistral displacement on the fault of about 10 km. The trace of the fault south-west of Gleann nan Eun is obscured by superficial deposits.

Other faults

Numerous minor faults occur throughout the Invermoriston district. Generally their presence is indicated by a reddening of bedrock and gullying due to the more friable nature of the faulted rock. It is possible to trace most of the faults by means of lineaments visible on air photographs and, although there is great variation, many trend approximately north-eastwards (Figure 33). It is sometimes possible to recognise the horizontal offset of steeply inclined geological boundaries but any vertical component of movement is usually indeterminate.

A fault north of Balnacarn (F) is less steeply inclined than most, dipping at approximately 40° to the ENE, and also has an exceptionally broad zone of shattering and reddening on the hanging wall side.

An example of a Permo-Carboniferous dyke occupying a fault (G) occurs in the Allt Dail a Chùirn [3065 0594]. The dyke has been intruded into a pre-existing fault and reactivation of the fault has led to localised crushing of the camptonite.

NINE

Quaternary

Ice streams emanating from the Northern Highlands glaciated the Invermoriston district during the Pleistocene period. Evidence from ice-moulded landforms, striae and erratics, notably of gneissose granite derived from the south-west, indicates a north-easterly movement of ice during the Devensian glaciation. This probably represents the main ice-movement direction throughout the Pleistocene. Erosion during the Devensian has removed much of the evidence of the earlier glaciations. Most landforms preserved today, particularly those within the Great Glen and Glen Moriston, are the result of events during the late-Devensian Loch Lomond Stadial.

MAIN LATE DEVENSIAN GLACIATION

Glacial striae are common on the surface of quartz veins, psammite and gneissose granite protected, until recently, by an overburden of glacial deposits. *Roches moutonées* and crag and tail features are common, particularly in large areas between Glen Urquhart and Glen Moriston. They are also found on the watershed between Glen Moriston and the Great Glen. These features form strongly linear arrays, especially where the strike of the foliation in the underlying bedrock coincides with the ice-movement direction, e.g. north of Loch ma Stac [34 22]. Larger-scale landforms related to the north-easterly ice flow include the streamlined shape of Mealfuarvonie [45 22], and the U-shaped valleys of Strathglass and the Great Glen. The form of Loch Ness is the result of glacial erosion of fractured rocks within the Great Glen Fault Zone.

Superficial deposits laid down during this glaciation are widespread. Sandy diamicton with lenses of sand and gravel commonly overlies dense compact grey-brown clayey tills. The thickness varies from more than 20 m in the valleys, to thin discontinuous veneers in the upland areas. Temporary exposures of till occur in the banks of many small streams as well as the major rivers. For example, the Allt Loch a' Cràthaich in Livishie Forest [3727 1882] incises into glacial deposits, more than 20 m in thickness, which rest on weathered psammite. Other occurrences of deeply weathered bedrock beneath substantial deposits of till occur along the watershed between the Great Glen and Glen Moriston, e.g. in the Allt Phocaichain between [3095 0995] and [3222 1057]. The exceptional thickness of these glacial deposits, and preservation of the weathered rock, might suggest the presence of a pre- Devensian valley that escaped erosion.

Meltwater from the retreating ice-sheet deposited large amounts of bedded sand and gravel. These commonly form the terraces and flat-topped mounds seen in Glen Urquhart, Glen Moriston and the Portclair Forest area, e.g. in the Allt a' Mhullain at [392 132]. Meltwater channels are common to the south-east of the River Enrick between [36 25] and [38 27].

Loch Lomond Stadial

Palynological evidence from Loch Tarff [425 100] suggests that milder climatic conditions prevailed during the Late-glacial interstadial (Pennington at al., 1972). Climatic deterioration in the Loch Lomond Stadial led to the development of an ice field in the Western Highlands and major outlet glaciers (the Loch Lomond Re-advance). Two of these glaciers flowed along the Great Glen terminating at Fort Augustus, and along Glen Moriston as far as the Dundreggan Reservoir [348 147]. Periglacial processes, including the development of regolith with stone stripes, were active beyond the ice limit.

GREAT GLEN GLACIER

In the Fort Augustus area (Figure 34) an assemblage of landforms, including hummocky deposits, outwash terrace gravels and raised shoreline fragments, marks the termination of the Great Glen Glacier (Sissons, 1979; Firth, 1984; 1993). On the western side of the Great Glen, the maximum extent of the Loch Lomond Stadial ice forms a recognisable feature from the shore of Loch Ness at Bunoich, to a height of 530 m in Coire a' Bhainne [2700 0668], 12 km to the south-west. In the upland areas a concentration of boulders, or a bench in the till cover, marks the limit. The re-advance limit between the Auchteraw Burn [3440 0890] and the Allt Dail a' Chuirn [3000 0678], and on Meall nan Calman [276 067], is a complex network of hummocky deposits, kames and meltwater channels. Between Jenkins Park [370 096] and Inchnacardoch [378 099], the hummocky deposits and *roches moutonées* are bounded to the north by a broad meltwater channel. Two prominent esker ridges of poorly sorted sand and gravel near Dail-a-chùirn [3085 0635] and Auchteraw [3445 0870] lie within, but at a high angle, to the ice limit (Sissons, 1979).

On the eastern side of the Great Glen above Fort Augustus, there is a distinct limit to the hummocky deposits. This rises from the shore of Loch Ness [3885 0870] along the Allt an Dubhair, to Tomamhoid [3867 0782] at an altitude of 100 m above OD. The limit extends south-westwards as a weak feature to Ardachy Wood [383 073], but is indistinct beyond Glen Tarff.

The form of the Great Glen within the re-advance limit is asymmetrical. The south-eastern flank is a steep, almost continuous, wall. A breach occurs only at Culachy [373 071] west of the River Tarff. A steep-sided

Figure 34
Geomorphology
of the Fort
Augustus area
(following Firth,
1984).

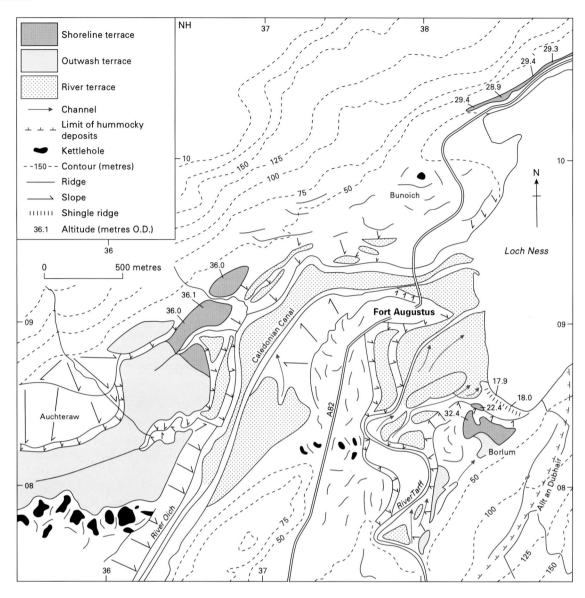

bluff marks the north-west side of the glen between Bridge of Oich [337 036] and Torr Dhùin [349 370]. This rises to a broad bench-like area, up to 3 km wide, between Lòn Mór [329 068] and Loch Lundie [296 037]. Linear NE-trending ridges of rock and fluted glacial deposits are prominent in this area. Substantial thicknesses of peat occur within the intervening depressions, e.g. Loch na Curra [322 039] and Loch Lundie [290 038]. Flat-bottomed drainage channels at Meall Mór, and on the Invervigar Burn between [318 055] and [325 056], cut across these features.

A constriction occurs in the floor of the Great Glen at Bridge of Oich [337 036], where the valley is less than 400 m wide. North-eastwards it broadens into an area between Auchteraw and Borlum of outwash deposits and shoreline fragments. Between Bridge of Oich and Torr a' Choiltreich [373 083] there is a line of NE-trending *roches moutonées*, rock ridges and crag and tail features. Striae are common on the rock surfaces. South-east of these ridges is a narrow steep-sided channel. This passes

north-eastwards into outwash channel deposits at the confluence with Glen Tarff. The channel is mostly infilled with peat, but sandy gravels occur near Coiltry [3605 0615].

At Borlum, many Late Devensian features are present (Figure 34). Ice-decay features characterise the area to the east of the River Tarff. Kame terraces to the south of Ardachy Lodge [380 071] lead into a meltwater channel through kame and kettled topography. This grades into a terrace fragment of coarse outwash gravel and boulder gravel, with thin lenses of bedded sand, e.g. at [3804 0796]. At Borlum [383 084] outwash material grades into a high-level terrace at 32.4 m above OD consisting of moderately well-sorted coarse gravel and lenses of bedded sand; Firth (1984; 1993) suggests that this is a lacustrine shoreline fragment. An erosional bluff truncates the ice-decay topography to north and west. At its foot is a small deposit of bouldery gravel [383 085], lying at 22.4 m above OD. From Borlum Cottage [3865 9837] to [3835 0856], at 17.9 to 18.8 m above OD, a narrow

2–3 m-high ridge of gravel extends along the southern shore of Loch Ness. The ridge and the bluff feature represent lacustrine features, charting changes in the level of Loch Ness (Firth, 1993).

The area between Torr Dhuin, Auchteraw and the River Oich contains substantial spreads of gravel deposits, in a kame and kettle topography (Figure 34). Rare sections near Auchteraw [3505 0730] and [3559 0730] reveal coarse, poorly sorted deposits of gravel and cobble gravel, more than 10 m thick. Lenses of pea gravel with a sandy matrix are common. These deposits are typical of high-energy fluvial deposition. Good examples of kettling are found at [355 077]. The limit of kettled gravel is approximate. The deposits grade into a discrete terrace flat of outwash gravel around Auchteraw (Sissons, 1979; Firth, 1993). Thin well-sorted gravel ridges within the surface of the terrace possibly represent channel deposits. North-eastwards from [365 089] and [367 092], the terrace passes into the well-sorted gravel, at 36 m above OD, of a high-level shoreline fragment (Firth, 1993).

Field evidence in the Loch Ness/Fort Augustus area suggests a sequence of events that spanned the period of the Loch Lomond Stadial (Firth, 1984; 1986). Along the northern shore of Loch Ness between Fort Augustus and Portclair [417 137], several distinct erosional bench fragments occur at an elevation of about 29 m above OD (Synge, 1977). Between Inchnacardoch [3820 1025] and the burn at [3902 1081], it forms a rock platform 5 to 20 m wide. Nearby [3913 1079] a broad terrace cut into the till occurs at a similar elevation. The terrace, some 110 m long and up to 70 m wide, is arcuate in form, with a small, steeply sloping back face locally preserved. The surface of both features slope gently towards the current shoreline of Loch Ness and are strewn with small rounded cobbles. Terrace fragments [3980 1150, 4180 1370] are relatively flat, the latter up to 80 m wide, and cut into the marginal till deposits. A concentration of orientated cobbles and larger blocks occurs at the back of these features, close to the break of slope (Synge and Smith, 1980).

These shoreline erosional features, indicate the main level of Loch Ness during the early part of the stadial. At the maximum extent of the re-advance glacier, the height of the loch stood at 32 m above OD. A temporary rise in the level to 36 m above OD occurred during the retreat of the ice-front. The rise in the water level and deposition of the large outwash spreads are considered (Sissons, 1979; 1981; Firth, 1984) to be consequential on the catastrophic drainage of the ice-dammed lake in Glen Spean, in the Glen Roy district (Sheet 63W). Erosion at the outlet of Loch Ness by the floodwaters subsequently lowered the loch to a new level of 22.5 m above OD. Lacustrine deposits are rare. However, a small outcrop of laminated clays occurs by the shore of Loch Ness

[382 095], at a height of about 21 m above OD. They comprise grey to buff grey, 1–2 mm horizontally laminated silts and clays, with thin intercalations of fine sand. Their relationship to the outwash deposits are unknown.

GLEN MORISTON GLACIER

The glacier that flowed eastwards along Glen Moriston dammed a lake in the upper part of Coire Dho in the Glen Affric district (Sheet 72E). On the southern slopes of Glen Moriston, the former limit of the glacier is marked by an elaborate system of end moraine ridges; the outermost ridge, followed for most of its length by a small meltwater channel, is the most continuous element (Figure 35). On the floor of the glen, kame and kettle topography, an esker up to 12 m high and pitted outwash terraces record the former ice tongue. The ice limit is suggested by the eastward termination of these features and by the absence of kettleholes from the lower, eastern, part of the outwash spreads (Sissons, 1977).

Extensive areas of water-washed bedrock occur on the north side of Glen Moriston (Figure 35). They are associated with interconnecting meltwater channels and kame terraces, and their upper limit records the westward rise of the former ice margin. Sissons (1977) considered the water-washed surfaces to be the result of sudden drainage of the ice-dammed lake in Coire Dho. Such drainage stripped away the former drift cover, often leaving a well-defined washing limit. The associated meltwater channels occur as two main types. The first occupy pre-existing valleys or depressions and have flat floors up to 0.35 km wide, infilled with glaciofluvial sediments commonly showing braided channel surfaces and terraces. These channels are associated with catastrophic lake drainage. The second are smaller, narrower channels typical of subglacial meltwater erosion. The channels and washed bedrock suggest a series of major floods following two major marginal, submarginal and subglacial drainage routes towards the snout of the glacier in Glen Moriston (Sissons, 1977).

HOLOCENE

The flood plains of the rivers Moriston, Glass and Enrick are underlain by alluvial deposits. Exposures in the river banks show that these comprise coarse gravel, commonly overlain by a metre or more of sand and silt.

Peat infills hollows and blankets gently inclined slopes over much of the district. To the north of the Great Glen Fault Zone, exceptionally extensive areas of thick peat occur around Loch ma Stac [33 21] and along the watershed between the Great Glen and Glen Moriston. Similar extensive deposits of peat are found on the western flanks of the Monadhliath Mountains, in the Glendoe Forest area [44 04].

Figure 35
Morphological
features
associated with
the Loch
Lomond
Readvance in
Glen Moriston
(following
Sissons, 1977).

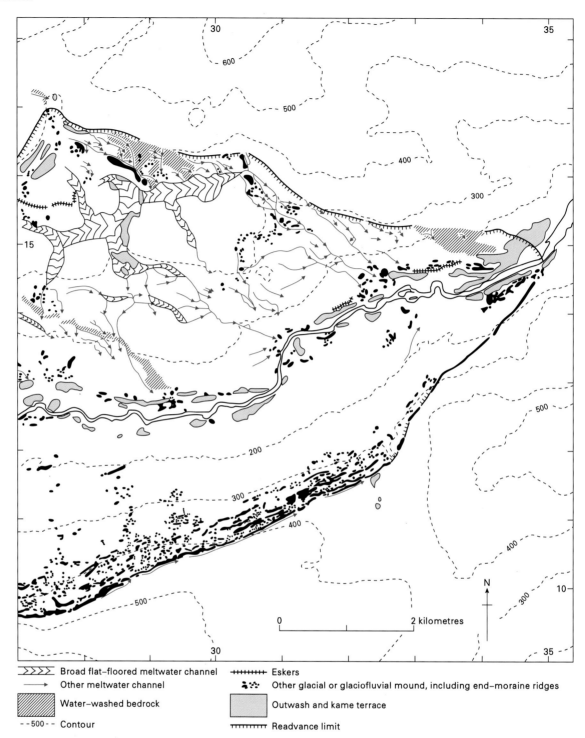

⊐⊐⊐⊐ Broad flat–floored meltwater channel	+++++++ Eskers
→ Other meltwater channel	⠿ Other glacial or glaciofluvial mound, including end–moraine ridges
▨ Water–washed bedrock	▨ Outwash and kame terrace
- -500- - Contour	⊓⊓⊓⊓ Readvance limit

TEN

Geophysics

The Lithospheric Seismic Profile in Britain (LISPB) geophysical line (Bamford et al., 1978; Barton, 1992) identifies three layers within the crust of the northern and central Highlands. The Neoproterozoic metamorphic rocks of the Caledonian orogenic belt (P velocity 6.1–6.2 km s^{-1}) overlie a pre-Caledonian basement (P > 6.4 km s^{-1}) which is underlain by lower crustal rocks (P ~ 7 km s^{-1}). The crust thickens from about 28 km at the Great Glen Fault, to about 35 km at the Highland Boundary Fault. The resolution of the seismic data is not sufficient to provide detail of the pre-Caledonian crust, or to show the effects of major structures such as the Great Glen Fault. However, magnetotelluric (MT) studies across the Great Glen (Kirkwood et al., 1981; Mbipom and Hutton, 1983; Meju, 1988) indicate the presence of broad-scale variations in conductivity at the deep crustal levels. The lower crustal rocks are relatively conductive (50–400 ohm m). In contrast, highly resistive upper crustal rocks occur on either side of the Great Glen Fault (2000–10 000 ohm m). A discrete, narrow (c.1 km) zone of conductive rocks at the fault extend into the lower crust. This, and other near vertical zones adjacent and parallel to the Great Glen, are tentatively interpreted as zones of major shear (Meju, 1988). There are no correlations across the fault within the Invermoriston district. Measurements of remnant magnetisation within rocks of the Foyers pluton (Torsvik, 1984) show much of the directed distribution is a present-day field component within multidomainal magnetite. However, older Siluro-Devonian components are present. In the Northern Highlands, these are indistinguishable from poles derived from granitic intrusions, giving a palaeo-pole of latitude 27° N and longitude 167° E (Torsvik, 1984; Storetvedt and Deutsch, 1986). Although the horizontal offset along the Great Glen Fault is not well constrained (Smith and Watson, 1983), substantial displacements since mid-Devonian times are unlikely (Briden et al., 1986).

BGS has not undertaken any ground geophysical surveys or deep borehole investigations within the Invermoriston district to constrain any crustal models. Some physical property data are available.

PHYSICAL PROPERTY DATA

Physical properties of the main rock types are presented in Table 13. Mean densities for the metasedimentary rocks include representative samples collected from outside the district. The magnetic susceptibility is a measure of the ease of magnetisation of a material which usually reflects its magnetite content. Values, expressed in SI 10^{-3} units, are derived from the logarithmic mean of up to 13 measurements per site. The most magnetic rocks are granodiorites from within the Foyers Plutonic Complex and magnetic units within the Glen Buck Pebbly Psammite Formation.

Density data show a significant contrast of about -0.06 Mgm^{-3} between psammitic and semipelitic lithologies within the Grampian Group. Lithologies within the Moine Supergroup have a mean density intermediate between psammite and semipelite. Granitic rocks show a wide variation in density values, with granodioritic rocks being significantly denser than monzogranitic lithologies.

MAGNETIC DATA

Aeromagnetic and gravity potential field data provide a important source of information on the nature of the regional crust. General studies exist of the Scottish Highlands (Hall, 1978), but are not specific to local geology. Aeromagnetic surveys across the Scottish Highlands were undertaken during 1962/63. Flown with a mean terrain clearance of 305 m, east–west flight lines were at

Table 13 Representative physical properties of rocks within and adjacent to the Invermoriston district.

Data source H BGS records.

Rock type	Saturated density (Mg m^{-3})			Magnetic susceptibility (SI 10^{-3})		
	Mean	No. sites	No. samples	Mean	No. sites	No. readings
Sedimentary rocks						
Devonian (ORS)	2.66	3	11	0.3	11	132
Metasedimentary rocks						
Grampian Group psammite	2.69	7	23	0.26	45	526
Grampian Group semipelite	2.75	2	9	0.95	14	169
Glen Buck Pebbly Psammite Formation	2.68	1	3	12.281	13	
Moine (undifferentiated)	2.73	2	9	0.40	14	205
Igneous rocks						
West Higland Granitic Gneiss	2.65	3	9	5.36	38	490
Foyers Plutonic Complex (granodiorite)	2.73	2	5	16.13	9	102

2 km intervals and north–south tie lines 10 km apart. The total field aeromagnetic anomaly map of the Invermoriston district and adjoining areas (Figure 36) derives from digital interpolation onto a 0.5 km grid. Contour values are of intervals at 50 nanotesla (nT).

There is a broad positive, nearly symmetrical anomaly centred along the Great Glen Fault Zone, with local maxima on the north-west side of the structure. The broad wavelength of the anomaly may represent magnetic basement, which becomes progressively shallower in the region of the fault. Local short-wavelength anomalies within the Northern Highland area, immediately to the south-west of the district, coincide with the outcrop of the Glen Garry Vein Complex. This probably reflects the presence of a buried plutonic mass. Similarly, the outcrop of the Foyers Plutonic Complex coincides with a major positive anomaly (300–400 nT) that is centred over the north-eastern part of the pluton (Kneen, 1973; Piper, 1979). This reflects the predominance of granodioritic rocks, which at outcrop have high magnetic susceptibilities (9–32 SI 10^{-3}). The monzogranitic rocks within the complex have smaller-amplitude magnetic susceptibilities (2–6 SI 10^{-3}).

To the north of the Great Glen Fault there is a broad decrease in the magnetic field. At outcrop the Moine lithologies have low magnetic susceptibilities (c.0.3 SI 10^{-3}). The numerous metabasic intrusions within the Moine outcrop and the Fort Augustus Granitic Gneiss have no apparent effect on the aeromagnetic anomaly pattern. This reflects the lack of magnetic minerals in the metamorphic mineral assemblages of these rocks.

To the south-east of the Great Glen Fault, a broad magnetic high remains after the removal of the shorter wavelength effects of the Foyers plutonic mass. This feature may represent a local magnetic basement. A planar gradient towards the Sronlairig Fault, to the south of the district, defines the south-east margin of this block. The north-east margin terminates abruptly as a steep NW-trending gradient (Figure 36), which may represent a trans-Caledonide deep crustal fracture. Rocks of the Grampian Group, like those of the Moine, have low magnetic susceptibilities (0.2–1.0 SI 10^{-3}). Prominent short-wave anomalies within the broad magnetic high to the south-east of the Great Glen Fault coincide with sheared rocks in the Glen Buck Pebbly Psammite Formation adjacent to the Eilrig Shear Zone. Susceptibility values of the sheared rocks are high (up to 40 SI 10^{-3}), although the formation as a whole is only moderately magnetic (c.12 SI 10^{-3}). Pebbly psammitic rocks in the adjacent Foyers district (Sheet 73E) show similar magnetic characteristics (Haselock and Leslie, 1992). As in the Northern Highlands, there is an overall decrease in the magnetic gradient away from the Great Glen. This may reflect a deepening of the magnetic basement. There is no positive evidence for the presence of Rhinns of Islay type crustal material.

GRAVITY DATA

The residual Bouguer gravity anomaly map (Figure 37) was compiled from data held in the National Gravity Data Bank. Values were derived using a reduction density of 2.72 Mg m^{-3} and interpolated on to a 0.5 km grid. A third-order polynomial surface was subtracted from the primary grid; this removed a broad regional gravity low, which originates from a deep crustal source across the Grampian Highlands, enhancing the upper crustal structures of the region. The derived complex anomaly pattern is poorly constrained, due to the wide distribution of the gravity data points and the paucity of density data.

Within the Northern Highlands, there is a gravity gradient which decreases away from the Great Glen. This broadly coincides with the regional magnetic high and suggests the presence of a dense magnetic basement that comes within 2 km of the surface at the Great Glen Fault. This basement can be modelled as having physical properties similar to Scourian granulites of the foreland sequence (Bott et al., 1972; Powell, 1978). Shorter-wavelength features are attributed to lateral density variations in the shallow crust, which may reflect the folding of the metasedimentary sequences.

To the south-east of the Great Glen Fault, the gravity gradients are complex. It is an interference pattern, deriving from the interplay of the gravity low of the Foyers plutonic mass and a substantial gravity high over this part of the Grampian terrane. The origin of this gravity high is uncertain.

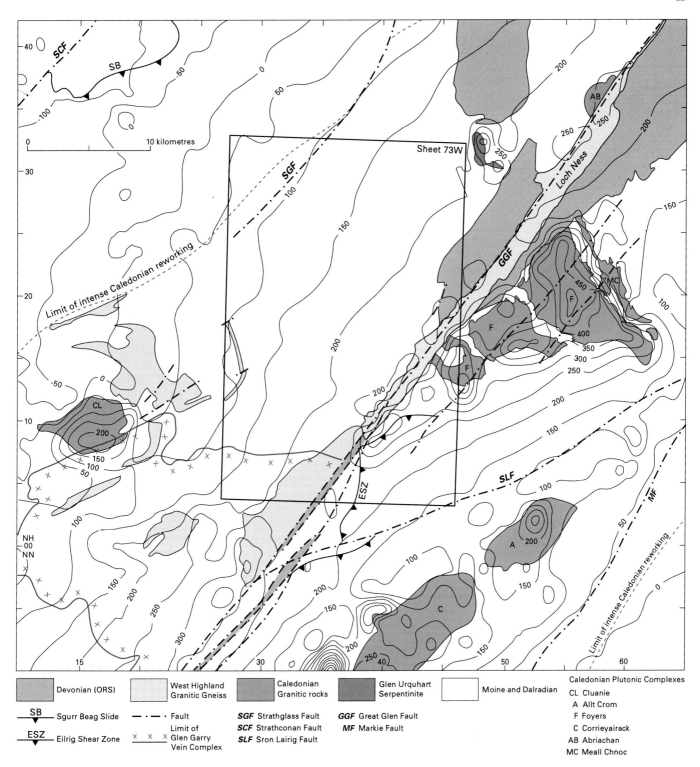

Figure 36 Total field aeromagnetic anomaly map of the Invermoriston district and adjoining areas. Contours are at 50 nanotesla (nT) intervals.

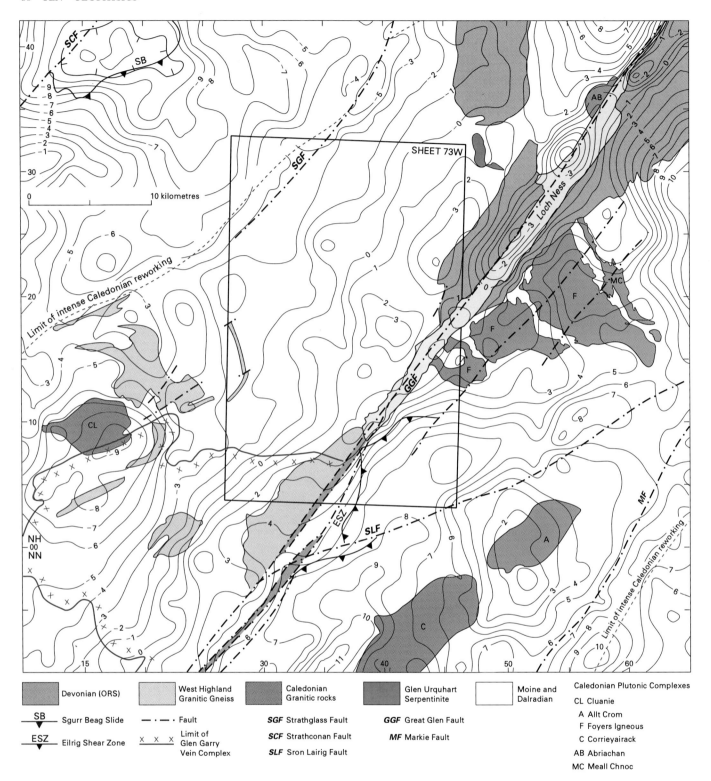

SHEET 73W

Figure 37   Residual Bouguer gravity anomaly map of the Invermoriston district and adjoining areas. Contours are at 1 mGal intervals.

	Devonian (ORS)		West Highland Granitic Gneiss		Caledonian Granitic rocks		Glen Urquhart Serpentinite		Moine and Dalradian

Caledonian Plutonic Complexes

SB — Sgurr Beag Slide

— · — · — Fault

SGF Strathglass Fault

GGF Great Glen Fault

CL Cluanie

A Allt Crom

F Foyers Igneous

ESZ — Eilrig Shear Zone

× × × Limit of Glen Garry Vein Complex

SCF Strathconan Fault

SLF Sron Lairig Fault

MF Markie Fault

C Corrieyairack

AB Abriachan

MC Meall Chnoc

ELEVEN

Economic geology

MINERALISATION

There is little evidence of mineralisation within the Invermoriston district. Minor vein mineralisation is found at two localities in Glen Moriston (Figure 33). In the Allt Bhlàraidh [3607 1771], quartz veinlets up to 0.2 m thick occur in a 3 m-wide zone, cutting rocks of the Upper Garry Psammite. The zone is subvertical, with an ENE trend. Quartz crystals are aligned normal to the vein margins. Crystal-lined cavities are common. The veins contain the following ore minerals: chalcopyrite, tetrahedrite, sphalerite, galena, pyrite and native silver. The silver and tetrahedrite occur as inclusions within the chalcopyrite. Alteration of the wallrock at the edges of the veins is generally extensive, with sericitisation common.

A quartz vein, in the Allt Iarairidh [3296 1710], containing calcite, pyrite, galena and sphalerite cuts the Upper Garry Psammite. The vein, up to 50 mm thick, dips at an angle of 63° to the south-east.

AGGREGATE

Large potential resources of both hard rock aggregate and sand and gravel exist within the district. However, there is an increasing requirement by the planning authorities to assess fully the environmental impact, servicing and of reclamation of any extractive developments.

Hard rock

The metasedimentary rocks of the district have historically provided a source of building stone, with the houses of many older settlements reflecting the local geology. Small abandoned workings occur in most lithologies. However, the facing stone of many larger, older buildings within the district, such as estate houses and hotels, is rarely of local origin. Triassic red sandstone is the most common. Stone for the construction of the hydroelectric dam at Dundreggan [357 157] largely derives from workings in the granodiorites of the Cluanie pluton in the adjacent Glen Affric district (Sheet 72E).

There are no active sites extracting hard rock within the district at present (1996). The remains of a large quarry are to be found to the north of Bridge of Oich [336 037] within the Fort Augustus Granitic Gneiss. This provided both dressed and armour stone during the construction of the Caledonian Canal. Plentiful resources of hard rock aggregate exist, with both the gneissose granite and metasedimentary psammitic lithologies providing potentially suitable sources of aggregate and armour stone.

Sand and gravel

Developments of glacial, glaciofluvial and fluvial sand and gravel deposits are extensive within the Great Glen and Glen Moriston. No assessment of the size and potential of the resource is available. The glaciofluvial deposits in Glen Moriston are of potential significance as a source of building sand and aggregate. To date, working of the resource satisfies only minor local needs.

Glaciofluvial and outwash deposits within the Great Glen between Torr Dhuin and Fort Augustus, and in the Borlum area, are extensive. Although the deposits are mostly coarsely clastic, they may prove a significant source of aggregate. Exploitation of the resource is minor and has mainly been for local purposes.

PEAT

Much of the Invermoriston district is peat-covered. Deposits in the lowland areas rarely exceed 1 m in thickness. There are, however, three significant developments in the upland areas, with recorded thicknesses greater than 3 m. These occur within the Glendoe Forest at the south-east corner of the district [43 06], on the watershed between the Great Glen and Glen Moriston [30 10], and in the catchment area of the River Enrick [31 21]. Historically, peat has been as a source of domestic fuel. Limited commercial extraction from these deposits might be possible, but is unlikely. Most areas of peat are coming under increasing pressure largely through drainage as part of both hill pasture improvement and forestry development.

REFERENCES

Most of the references listed below are held and are available for consultation in the libraries of the British Geological Survey at Murchison House, Edinburgh and Keyworth, Nottingham. Copies of the references can be purchased from the Keyworth Library subject to the current copyright legislation.

AFTALION, M, and VAN BREEMEN, O. 1980. U-Pb zircon, monazite and Rb-Sr whole rock systematics of granitic gneiss and psammitic to semipelitic host gneiss from Glenfinnan, north-western Scotland. *Contribution to Mineralogy and Petrology*, Vol. 72, 87–98.

ANDERSON, J G C. 1956. The Moinian and Dalradian rocks between Glen Roy and the Monadhliath Mountains, Invernessshire. *Transactions of the Royal Society of Edinburgh*, Vol. 63, 3–36.

ANDO, S. 1975. Minor element geochemistry from Mashu Volcano, eastern Kokaido. *Journal of the Faculty of Science, Hokaido University*, Vol. 16, 553–566.

BAMFORD, D, NUNN, K, PRODEHL, C, and JACOB, B. 1978. LISPB-1V. Crustal structure of northern Britain. *Geophysical Journal of the Royal Astronomical Society*, Vol. 54, 43–60.

BARR, D. 1983. Genesis and structural relations of Moine migmatites. Unpublished PhD thesis, University of Liverpool.

BARR, D. 1985. Migmatites in the Moines. 225–264 in *Migmatites*. ASHWORTH, J R (editor). (Glasgow: Blackie.)

BARR, D, ROBERTS, A M, HIGHTON, A J, PARSON, L M, and HARRIS, A L. 1985. Structural setting and geochronological significance of the West Highland Granitic Gneiss, a deformed early granite within Proterozoic, Moine rocks of north-west Scotland. *Journal of the Geological Society of London*, Vol. 142, 663–75.

BARROW, G. 1893. On an intrusion of biotite–muscovite gneiss in the south-east Highlands of Scotland and its accompanying metamorphism. *Quarterly Journal of the Geological Society of London*, Vol. 49, 330–358.

BARTON, P J. 1992. LISPB revisited: a new look under the Caledonides of northern Britain. *Geophysical Journal International*, Vol. 110, 371–391.

BOTT, M H P, HOLLAND, J G, STORRY, P G, and WATT, A B. 1972. Geophysical evidence concerning the structure of the Lewisian of Sutherland, north-west Scotland. *Journal of the Geological Society of London*, Vol. 128, 599–612.

BRIDEN, J C, TURNELL, H B, and WATTS, D R. 1986. Reappraisal of Scottish Ordovician palaeomagnetism: Reply to Storetvedt and Deutsch. *Geophysical Journal of the Royal Astronomical Society*, Vol. 87, 1207–1213.

BROOK, M, POWELL, D, and BREWER, M S. 1976. Grenville age for rocks in the Moine of north- western Scotland. *Nature, London*, Vol. 260, 515–517.

BROWN, G C, and LOCKE, C A. 1979. Space-time variations in British Caledonian granites: some geophysical correlations. *Earth and Planetary Science Letters*, Vol. 45, 69–79.

CAWTHORN, R G, STRONG, D F, and BROWN, P A. 1976. Origin of corundum-normative intrusive and extrusive magmas. *Nature, London*, Vol. 259, 102–104.

CHAPELL, B W, and WHITE, A J R. 1974. Two contrasting granite types. *Pacific Geology*, Vol. 8, 173–174.

CHAPELL, B W, and WYBORN, D. 1987. The importance of residual source material (restite) in granite petrogenesis. *Journal of Petrology*, Vol. 28, 1111–1138.

CHINNER, G A. 1966. The significance of aluminium silicates in metamorphism. *Earth Science Reviews*, Vol. 2, 111–126.

DEARNLEY, R. 1967. Metamorphism of minor intrusions associated with the Newer Granites of the Western Highlands of Scotland. *Scottish Journal of Geology*, Vol. 3, 449–457.

DEPAOLO, D J. 1981. Trace element and isotopic effects of combined wallrock assimilation and fractional crystallisation. *Earth and Planetary Science Letters*, Vol. 53, 189–202.

DICKINSON, W R, and SUCZEK, C A. 1979. Plate tectonics and sandstone compositions. *Bulletin of the American Association of Petroleum Geologists*, Vol. 63, 2164–2182.

FETTES, D J, and MACDONALD, R. 1978. Glen Garry vein complex. *Scottish Journal of Geology*, Vol. 14, 335–358.

FETTES, D J, and 6 others. 1985. Grade and time of metamorphism in the Caledonide Orogen of Britain and Ireland. 41–53 *in* The nature and timing of orogenic activity in the Caledonian rocks of the British Isles. HARRIS, A L (editor). *Memoir of the Geological Society of London*, No. 9.

FIRTH, C R. 1984. Raised shorelines and ice limits in the inner Moray Firth and Loch Ness areas, Scotland. Unpublished PhD thesis, Coventry Polytechnic.

FIRTH, C R. 1986. Isostatic depression during the Loch Lomond Stadial: preliminary evidence from the Great Glen, northern Scotland. *Quaternary Newsletter*, No. 48, 1–9.

FIRTH, C R. 1993. Fort Augustus. 192–196 in *Quaternary of Scotland*. GORDON, J E, and SUTHERLAND, D G (editors). (London: Chapman and Hall.)

GANGULLY, J, and SAXENA, S K. 1984. Mixing properties of alumino-silicate garnets: Constraints from natural and experimental data, and applications to geothermobarometry. *American Mineralogist*, Vol. 69, 88–97.

GRAHAM, C M. 1976. Petrochemistry and tectonic significance of Dalradian metabasaltic rocks of the SW Scottish Highlands. *Journal of the Geological Society of London*, Vol. 132, 61–84.

HALL, J. 1978. Geophysical constraints on crustal structure in the Dalradian region of Scotland. *Journal of the Geological Society of London*, Vol. 142, 149–155.

HARRIS, A L, PARSON, L M, HIGHTON, A J, and SMITH, D I. 1981. New/Old Moine relationships between Fort Augustus and Inverness. *Journal of Structural Geology*, Vol. 3, 187–188.

HASELOCK, P J, and LESLIE, A G. 1992. Polyphase deformation in Grampian Group rocks of the Monadhliath defined by a ground magnetic survey. *Scottish Journal of Geology*, Vol. 28, 81–87.

HASELOCK, P J, and WINCHESTER, J A, and WHITTLES, K H. 1982. The stratigraphy and structure of the Southern Monadhliath Mountains between Loch Killin and Glen Roy. *Scottish Journal of Geology*, Vol. 18, 275–290.

HIGHTON, A J. 1986. Caledonian and Pre-Caledonian events in the Moine south of the Great Glen Fault. Unpublished PhD thesis, University of Liverpool.

HIGHTON, A J. 1994. A re-evaluation of 'metasedimentary xenoliths' in the West Highland Granitic Gneiss of Inverness-shire. *Scottish Journal of Geology*, Vol. 30, 39–49.

HODGES, K V, and CROWLEY, P D. 1985. Error estimation and empirical geothermobarometry for pelitic systems. *American Mineralogist*, Vol. 70, 702–709.

HOLDSWORTH, R E, STRACHAN, R A, and HARRIS, A L. 1994. Precambrian rocks in northern Scotland east of the Moine Thrust: the Moine Supergroup. 23–32 *in* A revised correlation of Precambrian rocks in the British Isles. GIBBONS, W and HARRIS, A L (editors). *Special Report of the Geological Society of London*, No. 22.

IRVINE, T N, and BARAGER, W R A. 1971. A guide to the geochemical classification of the common volcanic rocks. *Canadian Journal of Earth Science*, Vol. 8, 523–548.

JOHNSON, M R W, and DALZIEL, I W D. 1966. Metamorphosed lamprophyres and the late thermal history of the Moines. *Geological Magazine*, Vol. 103, 240–249.

JOHNSTONE, G S, SMITH D I, and HARRIS, A L. 1969. The Moinian assemblage of Scotland. 159–180 *in* North Atlantic Geology and Continental Drift: a symposium. KAY, M (editor). *Memoirs of the American Association of Petroleum Geologists*, Vol. 12.

JOHNSTONE, G S, and MYKURA, W. 1989. *British regional geology: the Northern Highlands* (4th edition, revised). (London: HMSO for British Geological Survey.)

KENNEDY, W Q. 1949. Zones of progressive regional metamorphism in the Moine Schists of the Western Highlands of Scotland. *Geological Magazine*, Vol. 86, 43–56.

KEY, R M, MAY, F, and PHILLIPS, E R. 1991. Progressive deformation in part of the south-western Scottish Caledonides. *Terra Abstracts*, Vol. 3, 20/1.

KEY, R M, MAY, F, CLARK, G C, PEACOCK, J D, and PHILLIPS, E R. 1997. Geology of the Glen Roy district. *Memoir of the British Geological Survey*, Sheet 63E (Scotland).

KIRKPATRICK, R J. 1981. Kinetics of crystallisation of igneous rocks. 321–398 *in* Kinetics of geochemical processes. LASAGA, A C, and KIRKPATRICK, R J (editors). *American Journal of Science Reviews in Mineralogy*, Vol. 8.

KIRKWOOD, S C, HUTTON, V R S, and SIK, J. 1981. A geomagnetic study of the Great Glen Fault. *Geophysical Journal of the Royal Astronomical Society*, Vol. 66, 481–490.

KNEEN, S J. 1973. The palaeomagnetism of the Foyers Plutonic Complex, Inverness-shire. *Geophysical Journal of the Royal Astronomical Society*, Vol. 132, 53–63.

LE MAITRE, R W. 1989. *A classification of igneous rocks and glossary of terms: recommendations of the International Union of Geological Sciences Subcommission on the Systematics of Igneous Rocks.* (Oxford: Blackwell Scientific Publications.)

LOFGREN, G E. 1974. An experimental study of plagioclase crystal morphology: isothermal crystallization. *American Journal of Science*, Vol. 274, 243–273.

LOFGREN, G E and GOOLEY, R. 1977. Simultaneous crystallisation of feldspar intergrowths from the melt. *American Mineralogist*, Vol. 62, 217–288.

MACDONALD, R, GOTTFRIED, D, FARRINGTON, M J, BROWN, F W, and SKINNER, N G. 1981. Geochemistry of a continental tholeiite suite: late Palaeozoic dolerite dykes of Scotland. *Transactions of the Royal Society of Edinburgh: Earth Sciences*, Vol. 72, 57–74.

MARSTON, R J. 1971. The Foyers granitic complex, Inverness-shire, Scotland. *Quarterly Journal of the Geological Society of London*, Vol. 126, 331–368.

MBIPOM, E W and HUTTON, V R S. 1983. Geoelectromagnetic measurements across the Moine Thrust and Great Glen in northern Scotland. *Geophysical Journal of the Royal Astronomical Society*, Vol. 74, 507–524.

MEJU, M A. 1988. The deep electric structure of the Great Glen Fault, Scotland. Unpublished PhD thesis, University of Edinburgh.

MIYASHIRO, A, SHIDO, F, and EWING, M. 1970. Crystallisation and differentiation in abyssal tholeiites and gabbros from mid-ocean ridges. *Earth and Planetary Science Letters*, Vol. 7, 361–365.

MOULD, D D C P. 1946. The geology of the Foyers granite and the surrounding district. *Geological Magazine*, Vol. 83, 249–265.

MYKURA, W, and OWENS, B. 1983. The Old Red Sandstone of the Mealfurvonie Outlier, west of Loch Ness, Inverness-shire. *Report of the Institute of Geological Sciences*, No. 83/7.

NEATHERY, T L. 1965. Paragonite pseudomorphs after kyanite from Turkey Heaven Mountain, Cleburne County, Alabama. *American Mineralogist*, Vol. 50, 718–723.

PANKHURST, R J. 1979. Isotopic and trace element evidence for the origin and evolution of Caledonian granites in the Scottish Highlands. 18–33 in *Origin of granite batholiths*. ATHERTON, M P, and TARNEY, J (editors). (Orpington: Shiva.)

PARSON, L M. 1982. The Precambrian and Caledonian geology of the ground near Fort Augustus, Inverness-shire. Unpublished PhD thesis, University of Liverpool.

PEACOCK, J D, MENDUM, J R, and FETTES, D J. 1992. Geology of the Glen Affric district. *Memoir of the British Geological Survey*, Sheet 72E (Scotland).

PEARCE, J A. 1983. Role of the sub-continental lithosphere in magma genesis at active continental margins. 230–249 in *Continental basalts and mantle xenoliths*. HAWKESWORTH, C J, and NORRY, M J (editors). (Orpington: Shiva.)

PEARCE, J A, HARRIS, N B W, and TINDLE, A G. 1984. Trace element discrimination diagams for the tectonic interpretation of granitic rocks. *Journal of Petrology*, Vol. 25, 956–983.

PEARCE, T H, GORMAN, B E, and BIRKETT, T C. 1974. The TiO_2-K_2O-P_2O_5 diagram: a method of discriminating between oceanic and non-oceanic basalts. *Earth and Planetary Science Letters*, Vol. 24, 419–426.

PENNINGTON, W, HAWORTH, E Y, BONNEY, A P, and LISHMAN, J P. 1972. Lake sediments in northern Scotland. *Philosophical Transactions of the Royal Society of London*, Vol. B264, 191–294.

PHILLIPS, E R. 1992. Mineralogy, petrology and metamorphic history of the Eilrig Shear Zone, Fort Augustus, Sheets 63W and 73W, Scotland. *British Geological Survey, Technical Report* WG/92/46.

PHILLIPS, E R and KEY, R M. 1992. Porphyroblast-fabric relationships: an example from the Appin Group in the Glen Roy area. *Scottish Journal of Geology*, Vol. 28, 89–101.

PHILLIPS, E R, CLARK, G C, and SMITH, D I. 1993. Mineralogy, petrology and microfabric analysis of the Eilrig Shear Zone, Fort Augustus. *Scottish Journal of Geology*, Vol. 29, 143–158.

PIPER, J D A. 1979. Aspects of Caledonian palaeomagnetism and their tectonic implications. *Earth and Planetary Science Letters*, Vol. 44, 176–179.

PLATTEN, I M, and MONEY, M S. 1987. Formation of late Caledonian subvolcanic breccia pipes at Cruachan Cruinn,

Grampian Highlands, Scotland. *Transactions of the Royal Society of Edinburgh*, Vol. 78, 85–103.

POWELL, D, BAIRD, A W, CHARNLEY, N R, and JORDAN, P. 1981. The metamorphic environment of the Sgurr Beag Slide; a major crustal displacement zone in Proterozoic Moine rocks of Scotland. *Journal of the Geological Society of London*, Vol. 138, 661–673.

POWELL, D W. 1978. Gravity and magnetic anomalies attribual to basement sources under northern Britain. 107–114 *in* Crustal evolution in north-west Britain. BOWES, D R and LEAKE, B E (editors). *Geological Journal Special Issue*, No. 10.

POWELL, R, and EVANS, J A. 1983. A new geobarometer for the assemblage biotite-muscovite-chlorite-quartz. *Journal of Metamorphic Geology*, Vol. 1, 331–336.

RICHARDSON, S W, and POWELL, R. 1976. Thermal causes of the Dalradian metamorphism in the central Highlands of Scotland. *Scottish Journal of Geology*, Vol. 12, 237–268.

ROBERTS, A M, and HARRIS, A L. 1983. The Loch Quoich Line — a limit of early Proterozoic crustal reworking in the Moine of the Northern Highlands of Scotland. *Journal of the Geological Society of London*, Vol. 140, 883–892.

ROBERTS, A M, STRACHAN, R A, HARRIS, A L, BARR, D, and HOLDSWORTH, R E. 1987. The Sgurr Beag nappe: A reassessment of the stratigraphy and structure of the Northern Highland Moine. *Bulletin of the Geological Society of America*, Vol. 98, 497–506.

ROCK, N M S. 1977. The nature and origin of lamprophyres, some definitions, distinctions and derivations. *Earth Science Reviews*, Vol. 13, 123–169.

ROCK, N M S. 1983. The Permo-Carboniferous camptonite-monchiquite dyke suite of the Scottish Highlands and Islands: distribution, field and petrological aspects. *Report of the Institute of Geological Sciences*, No. 82/14.

ROCK, N M S. 1984. New types of hornblendic rocks and prehnite-veining in the Moines west of the Great Glen, Inverness-shire. *Report of the Institute of Geological Sciences*, No. 83/8.

ROCK, N M S. 1991. *Lamprophyres*. (Glasgow: Blackie.)

ROCK, N M S, MacDONALD, R, WALKER, B H, MAY, F, PEACOCK, J D, and SCOTT, P. 1985. Intrusive metabasite belts within the Moine assemblage, west of Loch Ness, Scotland: evidence for metabasite modification by country rock interactions. *Journal of the Geological Society of London*, Vol. 142, 643–61.

ROGERS, G, and 5 others. 1995. U-Pb evidence for the absence of Grenvillian event in the Moine of NW Scotland. In *EUGS Abstracts. Terra Nova*, Vol. 7, Abstract supplement 1, 353.

SANDERS, I S, VAN CALSTEREN, P W C, and HAWKESWORTH, C J. 1984. A Grenville Sm-Nd age for the Glenelg eclogite in north-west Scotland. *Nature, London*, Vol. 312, 439–40.

SAWKA, W N. 1988. REE and trace element variations in accessory minerals and hornblende from the strongly zoned McMurry Meadows Pluton, California. *Transactions of the Royal Society of Edinburgh: Earth Sciences*, Vol. 79, 157–168.

SHERVAIS, J W. 1982. Ti-V plots and petrogenesis of modern ophiolitic lavas. *Earth and Planetary Science Letters*, Vol. 59, 101–118.

SISSONS, J B. 1977. Former ice-dammed lakes in Glen Moriston, Inverness-shire, and their significance in upland Britain. *Transactions of the Institute of British Geographers*, Vol. 2, 224–242.

SISSONS, J B. 1979. Catastrophic lake drainage in Glen Spean and the Great Glen, Scotland. *Journal of the Geological Society of London*, Vol. 136, 215–224.

SISSONS, J B. 1981. Late glacial marine erosion and a jökulhlaup deposit in the Beauly Firth. *Scottish Journal of Geology*, Vol. 17, 7–19.

SMITH, D I. 1979. Caledonian minor intrusions of the N Highlands of Scotland. 683–697 *in* The Caledonides of the British Isles — reviewed. HARRIS, A L, HOLLAND, C H, and LEAKE, B E (editors). *Special Publication of the Geological Society of London*, No. 8.

SMITH, D I, and WATSON, J V. 1983. Scale and timing of movements on the Great Glen Fault, Scotland. *Geology*, Vol. 11, 523–526.

SOPER, N J and BROWN, P E. 1971. Relationship between metamorphism and migmatization in the northern part of the Moine Nappe. *Scottish Journal of Geology*, Vol. 7, 305–326.

STRACHAN, R A, and TRELOAR, P J. 1985. Discussion of 'A Grenville Sm/Nd age for the Glenelg eclogite'. *Nature, London* Vol. 314, 754.

STRACHAN, R A, MAY, F, and BARR, D. 1988. The Glenfinnan and Loch Eil divisions of the Moine Assemblage. 32–45 in *Later Proterozoic stratigraphy of the northern Atlantic regions*. WINCHESTER, J A (editor). (Glasgow: Blackie.)

STEPHENSON, D, and GOULD, D. 1995. *British regional geology: the Grampian Highlands* (4th edition). (London: HMSO for the British Geology Survey.)

STORETVEDT, K M and DEUTSCH, E R. 1986. Reappraisal of Scottish Ordovician palaeomagnetism. *Geophysical Journal of the Royal Astronomical Society*, Vol. 87, 1193–1206.

SWANSON, S E. 1977. Relation of nucleation and crystal-growth rate to the development of granitic textures. *American Mineralogist*, Vol. 62, 966–978.

SYNGE, F M. 1977. Land and sea level change during the waning stages of the last regional ice sheet in the vicinity of Inverness. 83–102 in *The Moray Firth area geological studies*. GILL, G (editor). (Inverness: Inverness Field Club.)

SYNGE, F M. and SMITH, J S. 1980. A field guide to the Inverness area. (Aberdeen: Quaternary Research Association.)

TALBOT, C J. 1982. Oblique foliated dykes as deformed single layers. *Bulletin of the Geological Society of America*, Vol. 93, 450–460.

TALBOT, C J. 1983. Microdiorite sheet intrusions as incompetent time and strain markers in the Moine assemblage north-west of the Great Glen fault, Scotland. *Transactions of the Royal Society of Edinburgh*, Vol. 74, 137–152.

TANNER, P W G. 1976. Progressive regional metamorphism of thin calcareous bands from the Moinian rocks of NW Scotland. *Journal of Petrology*, Vol. 17, 100–134.

TOBISCH, O. 1963. The structure and metamorphism of Moinian rocks in the Glen Cannich–Fasnakyle Forest area, Inverness. Unpublished PhD thesis, University of London.

TOBISCH, O. 1966. Large-scale basin- and dome-pattern resulting from the interference of major folds. *Bulletin of the Geological Society of America*, Vol. 77, 393–408.

TOBISCH, O. 1967. The influence of early structures on the orientation of late-phase folds in an area of repeated deformation. *Journal of Geology*, Vol. 75, 554–564.

TORSVIK, T H. 1984. Palaeomagnetism of the Foyers and Stontian granites, Scotland. *Physics of the Earth and Planetary Interiors*, Vol. 36, 163–177.

TYLER, I M. 1981. The metamorphic environment of the Strontian, Foyers and Dalbeattie intrusions, Scotland. Unpublished PhD thesis, University of Aston.

TYLER, I M and ASHWORTH, J R. 1983. The metamorphic environment of the Foyers Granitic Complex. *Scottish Journal of Geology*, Vol. 19, 271–286.

VAN BREEMEN, O, AFTALION, M, PANKHURST, R J, and RICHARDSON, S W. 1979. Age of the Glen Dessary syenite, Inverness-shire: diachronous Palaeozoic metamorphism across the Great Glen. *Scottish Journal of Geology*, Vol. 15, 49–62.

WATSON, E B, and HARRISON, T M. 1983. Zircon saturation revisited: temperature and composition effects in a variety of crustal magma types. *Earth and Planetary Science Letters*, Vol. 64, 295–304.

WELLS, P R A. 1979. P-T conditions in the Moines of the Central Highlands, Scotland. *Journal of the Geological Society of London*, Vol. 136, 663–671.

WHITTLES, K H. 1981. The geology and geochemistry of the area west of Loch Killin, Inverness-shire. Unpublished PhD thesis, University of Keele.

WINCHESTER, J A. 1974. The zonal pattern of regional metamorphism in the Scottish Caledonides. *Journal of the Geological Society of London*, Vol. 130, 509–524.

WINCHESTER, J A. 1976. Different Moinian amphibolite suites in northern Ross-shire. *Scottish Journal of Geology*, Vol. 12, 187–204.

WINCHESTER, J A and FLOYD, P A. 1976. Geochemical magma-type discrimination: application to altered and metamorphosed basic igneous rocks. *Earth and Planetary Science Letters*, Vol. 28, 459–469.

WINKLER, H G F. 1979. *Petrogenesis of metamorphic rocks* (5th edition). (New York: Springer-Verlag.)

YARDLEY, B W D, and BALTZAIS, E. 1985. Retrogression of staurolite schists and the source of infiltrating fluids during metamorphism. *Contributions to Mineralogy and Petrology*, Vol. 89, 59–68.

APPENDIX

Data sources

a. 1:10 000 and 1:10 560 maps (Solid and Drift).

The maps at 1:10 000 and 1:10 560 scales covering, wholly or in part, 1:50 000 Sheet 73W are listed with the names of the surveyors (D J Fettes, A J Highton, F May, J R Mendum, W Mykura, J D Peacock, N M S Rock, C G Smith, D I Smith) and the date of the survey.

Most of these maps are published and available for consultation in the Library, British Geological Survey, Murchison House, West Mains Road, Edinburgh EH9 3LA. Dyeline copies can be purchased from the Sales Desk.

NATIONAL GRID SHEETS

NH 20 NE	AJH, FM	1982–85
NH 20 SE	AJH	1984–85
NH 21 NE	DJF, FM	1982
NH 21 SE	FM	1982–85
NH 22 NE	JDP, NMSR	1972–78
NH 22 SE	DJF, JDP	1978–79
NH 23 SE	NMSR	1977–78
NH 30 NW	AJH	1982–83
NH 30 NE	AJH, DIS	1983–85
NH 30 SW	AJH, DIS	1979–85
NH 30 SE	DIS	1979–85
NH 31 NW	FM	1981–82
NH 31 NE	FM	1979–84
NH 31 SW	AJH, FM	1982–85
NH 31 SE	AJH, FM	1983–85
NH 32 NW	DJF, JDP, NMSR	1978–79
NH 32 NE	DJF, FM, JRM, CGS	1979–85
NH 32 SW	DJF, FM	1978–82
NH 32 SE	DJF, FM	1979–81
NH 33 SW	NMSR	1978
NH 33 SE	JRM	1978–79
NH 40 NW	DIS	1979–85
NH 40 NE	DIS	1979–85
NH 40 SW	DIS	1979–85
NH 40SE	DIS	1979–85
NH 41 NW	FM	1979–83
NH 41 NE	WM, DIS	1975–78
NH 41 SW	AJH, DIS	1985
NH 41 SE	DIS	1979–85
NH 42 NW	FM, CGS	1979–85
NH 42 NE	CGS	1981–82
NH 42 SW	FM	1979–85
NH 42SE	FM, WM	1975–85
NH 43SW	CGS	1978–79
NH 43 SE	CGS	1979–81

b. Geological Survey photographs

Twenty three photographs illustrating aspects of the geology of the Invermoriston district are deposited for reference in the library at the British Geological Survey, Murchison House, West Mains Road, Edinburgh. EH9 3LA. They belong to the D Series and were taken either in 1976, 1977, 1980 and 1995. The photographs depict details of the rocks and general views of the district. The photographs can be supplied, on request, as black and white or colour prints or transparencies, at a fixed tariff.

c. Petrography

Thin sections referred to in the text (e.g. S 71411), and many others collected during the survey but which are not specifically mentioned, are archived at the British Geological Survey, Murchison House, West Mains Road, Edinburgh EH9 3LA. They, and accompanying rock samples, together with rock samples collected and donated to the British British Geological Survey by Dr L M Parson, are available on short-term loan for further research.

INDEX

BRITISH GEOLOGICAL SURVEY

Keyworth, Nottingham NG12 5GG
0115 936 3100

Murchison House, West Mains Road, Edinburgh
EH9 3LA   0131-667 1000

London Information Office, Natural History Museum
Earth Galleries, Exhibition Road, London SW7 2DE
0171-589 4090

The full range of Survey publications is available through the Sales Desks at Keyworth and at Murchison House, Edinburgh, and in the BGS London Information Office in the Natural History Museum (Earth Galleries). The adjacent bookshop stocks the more popular books for sale over the counter. Most BGS books and reports can be bought from The Stationery Office and through Stationery Office agents and retailers. Maps are listed in the BGS Map Catalogue, and can be bought together with books and reports through BGS-approved stockists and agents as well as direct from BGS.

The British Geological Survey carries out the geological survey of Great Britain and Northern Ireland (the latter as an agency service for the government of Northern Ireland), and of the surrounding continental shelf, as well as its basic research projects. It also undertakes programmes of British technical aid in geology in developing countries as arranged by the Department for International Development and other agencies.

The British Geological Survey is a component body of the Natural Environment Research Council.

Published by The Stationery Office and available from:

The Publications Centre
(mail, telephone and fax orders only)
PO Box 276, London SW8 5DT
General enquiries 0171 873 0011
Telephone orders 0171 873 9090
Fax orders 0171 873 8200

The Stationery Office Bookshops
59–60 Holborn Viaduct, London EC1A 2FD
temporary until mid 1998
(counter service and fax orders only)
Fax 0171 831 1326
68–69 Bull Street, Birmingham B4 6AD
0121 236 9696   Fax 0121 236 9699
33 Wine Street, Bristol BS1 2BQ
0117 9264306 Fax 0117 9294515
9–21 Princess Street, Manchester M60 8AS
0161 834 7201 Fax 0161 833 0634
16 Arthur Street, Belfast BT1 4GD
01232 238451 Fax 01232 235401
The Stationery Office Oriel Bookshop
The Friary, Cardiff CF1 4AA
01222 395548 Fax 01222 384347
71 Lothian Road, Edinburgh EH3 9AZ
(counter service only)

Customers in Scotland may
mail, telephone or fax their orders to:
Scottish Publications Sales
South Gyle Crescent, Edinburgh EH12 9EB
0131 622 7050 Fax 0131 622 7017

The Stationery Office's Accredited Agents
(see Yellow Pages)

and through good booksellers